Evolution and the Myth of Creationism

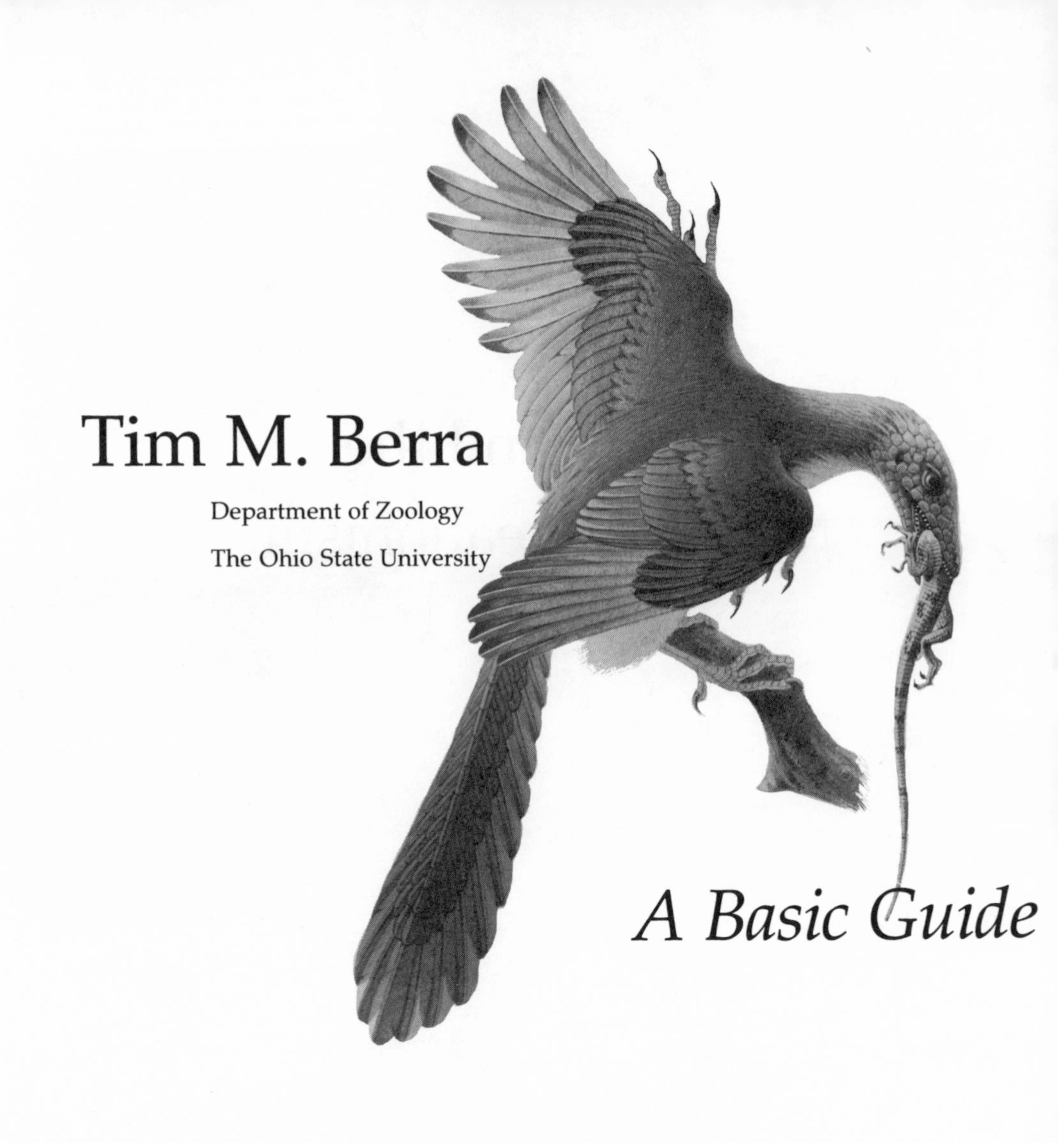

Tim M. Berra

Department of Zoology

The Ohio State University

A Basic Guide

EVOLUTION
and the Myth of
CREATIONISM

to the Facts in the Evolution Debate

To Joe Nelson in exchange
for the nifty little blue bag.
With Best Wishes
Tim Berra
6 July 2005
ASIH TAMPA

Stanford University Press, Stanford, California

Stanford University Press
Stanford, California
© 1990 by the Board of Trustees
of the Leland Stanford Junior University
Printed in the United States of America
CIP data appear at the end of the book

FOR MY MOTHER,

who allowed me to read during meals

Preface

I was originally drawn into the evolution/creation controversy in 1982, when I reviewed a draft of the Biology Curriculum Guide of the Columbus, Ohio, public schools. Until then, I had shared the view of most scientists that the creationists were not to be taken seriously. "Just ignore them, and their demands will soon be forgotten," was the attitude of busy scientists who didn't want to interrupt their research to do battle with the quaint notions of a bygone age—any more than astronomers should have to trouble themselves to refute the outlandish claims of astrology.

But astrologers are not campaigning to get sophistry taught as science in the public schools, and I was shocked to see that this Biology Curriculum Guide, a handbook for high school biology teachers, was about 50 percent creationist. It considered such fundamentalist Christian beliefs as that the Earth is only a few thousand years old to be the scientific equal of modern radiometric-dating techniques. Along the way, it trotted out the usual creationist chestnuts—misrepresentation of the Second Law of Thermodynamics, distortion of the fossil record, the claim of worldwide flooding, and many other examples of anti-science—passing them all off as science. The last straw came when, on pages 133 and 137 of the handbook, the *National Enquirer* was quoted as a scientific reference.

That was when I realized that science education is in trouble! It was clear that the strategy of ignoring the creationists would work to the detriment of both science and society. Apparently, many biologists and other scientists began waking up to the threat at about the same time, and scientists are now writing and speaking out to inform the public that creationism has no scientific validity and is a threat to the growth and spread of knowledge. (See the second reading list following the text for recent books inspired by the creationist attack on science.) To allow disguised, fundamentalist beliefs into the science classroom is to devalue education at a time when our country needs the best scientific minds it can muster just to stay even with the worldwide technological explosion. To teach students that the foundations of biology, most of geology and astronomy, and a good deal of physics are flawed is to cheat them, shackle their intellectual growth, erode their ability to compete for jobs, and stifle their prospects for a rewarding life.

Creationists, for the most part, are fundamentalist Christians whose central premise is a literal interpretation of the Bible and a belief in its inerrancy. In adopting a literal interpretation of the Bible, they differ from nearly all other Christians and Jews. Scientists, many of whom are religious, have no wish to deny fundamentalists their own beliefs, but the creationists are determined to impose their views on others. In particular, they are lobbying to have science classes teach the ideas of: a sudden creation from nothing by God; a worldwide flood; a young Earth; and the separate ancestry of humans and apes. These ideas constitute the biblical story of creation and, as such, are inherently religious. And because they depend on supernatural intervention, not natural law, they are also unscientific. There is no scientific evidence, or even an appeal from common sense or experience, to support the creationists' claims. The few claims that are susceptible to testing, such as a young Earth, are shown by the scientific evidence to be false. Yet creationists do not remove demonstrably false ideas from their pantheon of beliefs. Unlike scientists, they do not subject their assertions to revision based on evidence. Since this pattern of thought is not scientific—does not require or even condone testing and retesting—it should not be taught in science classrooms in public schools.

The creationists are determined to force their will on society and the schools, through the courts if possible. Their strategy—

ironically enough, considering the moral precepts of Christianity —is founded in deception, misrepresentation, and obfuscation designed to dupe the public into thinking that there is a genuine scientific controversy about the validity of evolution. *No such controversy exists*, but it is difficult for the lay public to distinguish the scientists, who often disagree on the nuances of evolutionary theory (but not on evolution's existence), from the creationists, who stick together and cloak absurd claims in scientific terminology. Creationist arguments are discussed in Chapter 5 of this book, and the reader is also referred to *Science* (1982, 215: 934–43) for the legal opinion of U.S. District Court Judge William R. Overton on the creationists' claims to be "scientific."

This book has three related purposes. First, it is an attempt to explain evolution to people who are genuinely confused by the claims of creationists, who try to present fundamentalist Christian beliefs as science. Second, it seeks to provide useful ammunition to the high school biology teacher or school board member who finds himself or herself under attack by creationists. Third, it should be a useful supplemental text for introductory college-level classes in biology, zoology, botany, or anthropology. These three purposes can all be served by answering the following question: What should an educated person know about the theory of evolution? I have tried to provide that information in this little book.

I am writing for the open-minded reader who does not understand the technical issues of evolution, but would like to, who sees everywhere the signs of a bitter philosophical and educational debate, but does not know what to make of it, or who to believe. Those readers who demand a literal interpretation of the Bible will probably not be swayed by this book, for they have chosen to abandon reason and evidence in favor of dogma and blind faith. Several scientist colleagues who read this book in one of its sixteen drafts admonished me to tone down the bluntness of my statements on creationism, fearing that I might alienate otherwise receptive readers. But it seems to me that scientists have, for too long, treaded too lightly on the creationists, and have thereby fostered the impression that the creationists are a legitimate scientific voice. It is time for candor and clarity.

Much in modern life has become dismayingly technical, and

evolutionary theory is no exception, but a generation that has learned to cope with VCR's, word processors, and instant teller machines should be able to handle the basic facts of evolution. The terms given in boldface type in the text are defined in the Glossary at the end of the text, but these terms are also generally explained where they first occur.

The manuscript for this book benefited greatly from detailed criticism by Dr. Ward Watt, Department of Biological Sciences, Stanford University; Dr. Robert C. Cashner, Department of Biological Sciences, University of New Orleans; Dr. Eugenie C. Scott, Executive Director, The National Center for Science Education; Dr. John A. Moore, Professor of Biology, University of California, Riverside; Dr. John Lynch, Department of Life Science, University of Nebraska; Dr. Warren D. Dolphin, Department of Zoology, Iowa State University; Dr. David B. Wilson, Departments of History and Mechanical Engineering, Iowa State University; and five anonymous reviewers. I am pleased and grateful that they took their assignment so seriously. Further suggestions for improvement came from Dr. Sidney Fox, University of Miami, Dr. Michael Ruse, University of Guelph, and Dr. Donald C. Johanson, Director of the Institute of Human Origins, Berkeley. The comments of my introductory biology students helped me decide what material to include. The remarks of my scientific colleagues at Ohio State University—Dr. Abbot S. Gaunt, Professor Sandra L. Gaunt, Dr. Gene Poirier, Dr. Janet Tarino, Professor Louise Troutman, and Dr. Terri Fisher—as well as those of my nonscientific OSU colleagues, Professors Tony Pasquarello, Ralph Hunt, and Richard Wink; and my father-in-law, Jack Flynn, improved the readability of the manuscript. I deeply appreciate all this help. Stanford University Press editor William W. Carver was a veritable fountain of ideas that splashed over each and every page of this manuscript. I am also grateful for his generously shared wordsmithing, which helped keep my rhetoric and technical prose in check. Three student artists in the Medical Illustration Department of Ohio State University, Jeanette M. Burkhardt, Laura K. Ellerbrock, and Stephanie A. Sugar, under the supervision of their instructor, Richard Hall, undertook the illustration of this book as a class project; 26 of the illustrations are their work. David Dennis prepared two of the drawings in Appendix A. Babette Mullet cheerfully typed fourteen revisions of the evolving manuscript over a six-year period. Her early efforts were made possible by the word-processing tutelage of my colleague, the late Dr. David Scott. Marna Utz typed the final revisions. George Wenzel and Gail Bird made the task of indexing painless.

T.M.B.

Contents

Figures and Tables

Figures

Tables

Evolution and the Myth of Creationism

1 What Is Evolution?

How do scientific theories earn their place in textbooks? Why is evolution a scientific theory and creationism not science at all? What is evolution, and what evidence is there that it has occurred? These are the topics of this chapter.

Hypotheses and Experiments

Where Do Scientific Theories Come from?

When you walk into your bedroom and flick the light switch and nothing happens, what do you do next? Chances are, you will turn the switch on and off a few times. Even though you are not conscious of doing so, you have formed a **hypothesis** (an educated guess) that the switch isn't making proper contact, and you are performing an **experiment** (trying the switch again) to test that hypothesis. When the light still fails to glow, you reject the "bad-contact" hypothesis and replace it with the "bad-bulb" hypothesis. You test this by installing a new bulb. If the new bulb lights, you have confirmed the bad-bulb hypothesis, but if there is still no light, you check the fuse box or circuit breaker. We use these logical, commonsense steps many times each day without thinking about the process. *Scientists use these same steps consciously*; how they proceed is dauntingly termed the **scientific method**, but the method is not difficult to understand.

The vast body of knowledge generally known as "science"

proceeds via the scientific method. Activity that does not function this way is a poor bet to uncover truth, a poor way, for example, to design a bridge we would care to trust, and cannot be called science. One does not need to be a professional scientist to come up with a useful insight into the workings of the real world, but the *validity* of the insight can be demonstrated only by the patient, logical, objective workings of science. The public, confronted every day by the mysteries and symbols of higher mathematics, chemistry, genetics, and electronics, often seems to suspect that scientists are voodoo doctors, shrouding their secrets and rituals in obscure language so as to exclude ordinary citizens from their private camp. No doubt some scientists, the more insecure among them, do that, and a few—like people in any walk of life—cheat and lie, too. But in fact *there is no domain of human knowledge or endeavor that is more open to scrutiny than science*; it is in the very nature of science that it be honest, fair, and aboveboard, ready at all times to admit its errors and revise its theories, and when scientists are caught faking their laboratory results, in support of a doubtful hypothesis, they know they have bought their careers a one-way ticket to oblivion. Without these checks on its practices, science would be doomed to failure: serious researchers would be few and beleagured, and we would have no polio vaccine, no space flight, no television, no computers, not even plastic garbage bags.

The scientific method involves the observation of phenomena or events in the real world, the statement of a problem, some reflection and **deduction** on the observed facts and their possible causes and effects, the formation of a hypothesis, the testing of the hypothesis (**experimentation** or **prediction**), and—when the tests repeatedly confirm the hypothesis—the erection of a **theory**. First, a problem is identified from observations (for example, why does x always seem to do y when z is present?). A hypothesis is then formed to resolve the problem. To test the hypothesis, experiments are performed or predictions are made. If, after repeated testing using different systems, equipment, organisms, etc., by large numbers of scientists over days or weeks or years, the hypothesis has not been refuted and continues to explain the hard data, then that hypothesis is accorded the status of a theory of science. *If, however, the hypothesis fails the tests, it is discarded.* Most hypotheses are in the end discarded, and only a few of these—victims, perhaps, of faulty data or poor testing the first time around—ever resurface as serious proposals later.

Thus, science "discovers" truth by a never-ending process of elimination; the single logical possibility still standing after careful scrutiny of all available data and all competing hypotheses becomes—for as long as it withstands new challenges—the theory upon which new research builds. This method is as old as Aristotle, and it works. No theory, even evolution, is ever held to be sacred; but some are more durable than others, as near to unchallengeable as makes little difference.

A scientific theory, then, in addition to explaining the data, has predictive value. Making testable predictions can replace experimentation in hypothesis testing in some cases. For example, comparative biologists called **systematists** study the relationships between organisms and arrange them into a **classification** that is intended to reflect their ancestry; it is not possible for systematists to study the genes or behavior of long-extinct animals, but they *can* make predictions about the fossils that are still to be discovered. The theory of evolution, developed and refined over more than a century by thousands of biologists, has been very useful in providing predictions that survive repeated testing. Thus, a scientific theory is more than an array of logically correct propositions. It is also the distillation of a collection of evidence that dependably describes some part of the real world.

A scientific theory must also be capable of being falsified. That is, there must be some set of circumstances such that, *if they were to occur*, the theory would fail to explain the facts. Scientists expend a great deal of time and effort devising tests of falsification, for any theory incapable of being falsified, such as that a God exists or that people have souls, is not a scientific one. On this basis, if mammalian fossils were found in rocks older than those that contain the first fish fossils, then a major element of the theory of evolution—that mammals arose far more recently than fishes, indeed ultimately *from* fishes through amphibians and reptiles—would be falsified. The fact that mammalian fossils occur only in rocks that are relatively a good deal younger than the earliest fish fossils supports the theory and protects it from rejection by that particular challenge. The distribution of animal groups over time, as revealed by the fossils in the ground, has been shown to be consistent with their hypothesized evolutionary origins.

Science, because it is a self-correcting endeavor, is prepared to modify any theory, no matter how cherished, if the data show that to be necessary. And indeed, many scientific theories that stood for decades have been abandoned when critical new data came to light. When a major, long-held theory at last gives way to a powerful and radically different new theory, we call that revolution a **paradigm** shift, and such a shift is usually followed by an explosion of new research and discovery.

But a theory in the special scientific sense is not "just a theory," as creationists are fond of saying. A *scientific* theory is the endpoint of the scientific method, often the foundation of an entire field of knowledge, and is not to be confused with the sort of "theory" we so easily propose in everyday conversation. For example, the theory of evolution is considerably more substantive, and has survived far more challenges, than the "conspiracy theory" of the assassination of President Kennedy.

Literal Interpretation and the Scientific Method
Must Reason and Faith Forever Conflict?

In contrast, the methods and claims of creationists are not subject to experimentation, prediction, revision, or falsification. To them, these pursuits are irrelevant, because they believe they possess the "truth" as set forth in the Bible. In their view, scientific theory or evidence that contradicts the Bible must be in error, because the Bible (actually, their particular interpretation of the Bible) cannot be mistaken. This view is clearly stated by creationist spokesman Henry M. Morris, Director of the Institute for Creation Research in San Diego, who asserted, "It is precisely because Biblical revelation is absolutely authoritative and perspicuous that the scientific facts, rightly interpreted, will give the same testimony as that of Scripture. There is not the slightest possibility that the *facts* [emphasis his] of science can contradict the Bible and, therefore, there is no need to fear that a truly scientific comparison of any aspect of the two models of origins can ever yield a verdict in favor of evolution" (Morris, 1974, *Scientific Creationism*, pp. 15–16). Elsewhere, Morris wrote, "The only way we can determine the true age of the earth is for God to tell us what it is. And since He *has* told us, very plainly, in the Holy Scriptures that it is several thousand years in age, and no more, that ought to settle all basic questions of terrestrial chronology" (Morris, 1978, p. 94). These views are most definitely not applications of the scientific method or even of common sense.

Thus the creationists must ignore or distort the evidence for evolution in order to force it into compliance with their interpretation of the Bible. For example, creationists *interpret* the Bible to be saying that the Earth was formed about 10,000 years ago. The vast array of scientific data pointing to an ancient Earth, drawn from nuclear physics in the form of radiometric dating, is discarded by the creationists because it doesn't fit their view. This is not a scientific approach, and the term "scientific creationism" is an oxymoron—a self-contradicting phrase. Most Christians and Jews are content to see the Bible as metaphor or mystery or allegory, at least in the particulars of the Book of Genesis, but the fundamentalists, the creationists, believe it can be (and must be) interpreted literally.

A literal interpretation of the scriptures poses problems for the fundamentalists because there are two quite different versions of creation in Genesis. The first story ("In the beginning . . .") starts at the beginning of the first chapter of Genesis and ends early in the second chapter. The second creation story (The Garden of Eden) begins in the second chapter and continues through the third. These two stories describe creation events in contradictory sequences and in different terms, and the theological messages are distinctly different. Which account do creationists claim is literally, historically true? The other must then be false or at least not taken literally. A thoughtful analysis of this fundamentalist dilemma is given by Beck (1982).

Natural Selection and Evolution
Is Life Really a War for Survival?

The steps of the scientific method, established long before Darwin, were followed very carefully in the development of evolutionary theory. There have been many ideas related to biological evolution in the history of human thought, but the only one to survive the test of time was proposed by the English naturalists Charles Darwin (1809–1882) and Alfred Russel Wallace (1823–1913) in separate papers read before the Linnean Society in London in 1858. In November 1859, Darwin (Figure 1) published *On the Origin of Species*, in which he not only elaborated the theory of evolution, but also proposed a mechanism by which it could work. Today the theory of evolution forms the foundation of the biological sciences and their applied subdisciplines of medicine and agriculture, by providing the conceptual framework for both experimentation and prediction.

Darwin noticed that many more offspring of various plants and animals were produced than actually survive. A female cod, for example, can produce over 1,000,000 eggs per year; a maple tree produces thousands of its winged fruit each spring; the oceans teem with larvae of all sorts that will never reach maturity. Thus the reproductive capacity of organisms greatly exceeds any actual, realized population size. Overproduction of offspring is a fact of nature, one that cries out for explanation. Darwin also noted that no two individuals of any **species**, except the very rare identical twins, are utterly alike. In other words, there is variation in nature, everywhere. Darwin therefore reasoned that there is **competition**—for survival, mates, space, food, shelter, and other resources—in which the favorable variations tend to be preserved by nature and the unfavorable ones tend to die out. He called this process **natural selection**. The consequence of natural selection is biological **evolution**, which Darwin termed "descent with modification." This is still, 131 years later, considered a good descriptive definition of evolution.

Darwin had no knowledge of **genetics** (the science of **heredity**); and the **fossil record**, although beginning to be known in 1859, was not nearly as well understood as it is now. Today we

Figure 1. Charles Darwin at the age of 54. This photo was taken in 1863, four years after the publication of *On the Origin of Species*. Photos of a bearded Darwin occur in 1866 and thereafter. (The original of this photograph is in the Candolle Collection at the Conservatoire et Jardin Botaniques de la Ville de Genève; reproduced here by courtesy of the Hunt Institute for Botanical Documentation.)

have the benefit of genetics and a more complete **paleontology** (the scientific study of fossils) in explaining evolution. It is a remarkable achievement that Darwin arrived at essentially the view that science has today without any genetics at all and with a somewhat limited view of the fossil record. (A chronology of Darwin's life and achievements follows the main text, as Appendix B.)

In today's terminology the relationship between natural selection and evolution is explained as follows. Some genetic variants may be better adapted to their environment than others of their sort, and will therefore tend to survive to maturity and to leave more offspring than will organisms with less favorable variations. This **differential reproduction** of genetic variants is the modern definition of natural selection. It results in a change in the frequency of occurrence of certain **genes** over time within a population—more of some genes, fewer of others. Changes of this sort, and their manifold consequences as the generations come and go, constitute the definition of evolution. In summary, *evolution is a change of gene frequency* brought about by natural selection (differential reproduction) and other processes acting upon the variations produced by **sexual reproduction**, **mutation**, and other mechanisms. The **environment** is the selecting agent, and because the environment changes over time and from one region to another, different variants will be selected under different environmental conditions. The Canada of the ice age was a different environment than the Canada of today.

The biological sciences, like astronomy or nuclear physics, have grown so rapidly, amassed so much data, and developed so many theories and subdisciplines that even biologists are unable to stay abreast of all their findings, or even of all their general lines of development. For the layperson, then, or the writer attempting to simplify things for the layperson, the challenge of evolution is formidable. I will try in what follows to simplify wherever possible, but the reader is advised to be patient, and to move on even when all is not understood—the general shape of things, at least, will usually be clear. Those readers who want a more technical explanation of the process of mutation and genetic variation can consult Appendix A and the Glossary, at the end of the text; the terms given in boldface type in the text are defined in the Glossary.

Nonadaptive Evolution
Is Change Always for the Better?

Natural selection is reflected particularly in **adaptation** whereby an organism becomes fine-tuned to its environment. The hummingbird's long, narrow bill, extrusible tongue, and hovering ability are exquisitely adapted for probing deep into the corollas of tubular flowers; the mole's spadelike hands are great digging tools; and so on. But although natural selection is the *major* source of evolutionary change, it is not the *only* one. Darwin was aware that other forces besides natural selection were involved in descent with modification, but to have appreciated the **nonadaptive** causes of evolution discussed below, he would have needed a knowledge of modern genetics. (Note, by the way, that nonadaptive should not be taken to mean anti-Darwinian.)

Evolutionary change is typically driven by environmental forces, but it may also be random or neutral. For example, suppose there is a population of snails composed partly of dark individuals, partly of light individuals, and suppose a hurricane blows one of the snails far away, depositing a dark, self-fertilizing **hermaphrodite** on an island (many snails are hermaphroditic). This single snail may eventually produce an entire **colony** of dark snails, particularly if no other species of snail already occupies the available **niches**. The gene frequency of the new population will be quite different from that of the ancestral population because the founder (our lone, dark snail) carried—as all individuals of a species do—an incomplete sample of the original population's **gene pool** (all of the genes of all individuals in the population). Owing to sheer chance, not natural selection, some advantageous genes may be lost in the process, and some formerly scarce genes may be common in the new population. This is called the **founder principle**, and it operates within very small, newly established populations, such as those animals and plants that first arrived and became established in the Galapagos

Islands, after volcanic action had thrust the islands above the ocean's surface.

Genetic drift is a similar phenomenon that results in the random loss of **alleles** (alternate forms of a given gene—blood type A vs. blood type B, for example). In a small population, certain genes, perhaps including favorable ones, can be eliminated by the wholly accidental death of their carriers, before they have reproduced. This sort of change in gene frequency is not a result of natural selection. For example, the only two toads with novel skin pigmentation in a population of drab individuals may be squashed by a beer truck while they are crossing a highway at night. Their death is not related to the survival value of their genes for novel skin pigmentation; it is just bad luck, not natural selection.

The founder principle and genetic drift can operate together to produce a change in gene frequency (which, again, is what evolution is) that is different from the change that would have been brought about by either process acting alone, and is not brought on by natural selection. A great deal of the new species formation that has occurred on islands has been attributed to these two processes, followed by the sifting of favorable variations through centuries and millennia of natural selection.

Another mechanism of nonadaptive evolution is **mutation pressure**, which involves a change of gene frequency due to the more frequent occurrence of a mutation than its corresponding **back mutation**. In other words, because of the chemical reactions involved, it may be easier for the chemical mistake in gene replication (mutation) to occur in one direction (from A to B) than in the opposite direction (from B to A). Even mildly harmful mutations that are ordinarily removed by natural selection can become established in a population if they arise at a rate faster than natural selection can remove them.

There is a fair amount of redundancy, or repetition, in the **genetic code**. A specific **amino acid** (amino acids are the building blocks of **proteins**) may be dictated by any one of several **codons** (the basic coding units of the DNA molecule; Figures A1–A5 and Table A1 in Appendix A provide detailed descriptions for those readers who want to understand how the genetic code influences **protein synthesis**). A mutation that converts one codon to another is undetectable by natural selection as long as the end product (the amino acid) is the same. This kind of change is a **neutral mutation**, and it can spread through a population by mutation pressure or genetic drift. Neutral mutations, which produce no outwardly observable effects, are thought to be much more common than previously suspected. They are responsible for much nonadaptive change, but they cannot account for adaptive change. Natural selection alone is responsible for adaptation.

These nonadaptive sources of evolution (founder principle, genetic drift, mutation pressure, and neutral mutation) demonstrate that Darwin did not have the last word on evolution. The incompleteness of his theory is easily understood in the context of nineteenth-century science, but he did point us in the right direction. We are still learning.

It is the process of evolution, led in part by natural selection and in part by the various nonadaptive causes of gene-frequency change mentioned above, that has produced the diversity of life on Earth. In a similar manner, humans direct the evolution of domestic crops and animals via **artificial selection**. Consider the great variety of dog breeds, the improved strains of crops and livestock, and so forth. All of these were developed by humans choosing which organisms would be allowed to contribute their genes to the next generation through selective breeding. Nature does the same thing—much more slowly.

Populations and Gene Flow
How Do New Species Evolve?

Evolution by natural selection and the various other mechanisms mentioned above may lead, over time, to slight changes or very large changes in the descendants of the original organisms. Biologists sometimes divide evolution into two processes: **microevolution**, or change in gene frequency within a population, which may lead to the formation of new species; and **macroevolution**, which involves evolutionary change above the species level, as for example in long-term trends within whole lineages, or in mass extinctions. The relationship between the two processes is currently a popular research topic. Most biologists think that macroevolution is simply the result of the accumulation of great numbers of the same sort of small changes that produce microevolution. Other biologists argue that the distinction runs deeper.

The word "evolution," in any case, is sometimes used imprecisely, to mean "speciation." **Speciation**, or the formation of new species, requires evolution but is not synonymous with it. Not all evolution leads to the formation of new species, since considerable evolution can occur, across time, *within* a given species, but speciation is a logical consequence of natural selection and genetics operating within an environmental context.

Most speciation requires two things: heritable variation and geographic isolation. Let us consider a population of land snails. (A **population** is a group of interbreeding individuals of the same species occupying a given area.) Each snail is capable of exchanging genes with every other snail in the population. Over geologic time a lake forms in the middle of the area occupied by the population, separating the snails into two groups, A and B (Figure 2). The snails cannot cross the lake. Let us assume also that the prevailing winds acting on the surface of the lake result in somewhat different climates in the two areas: area A becomes warm and dry; and area B becomes cool and damp. Thus the two snail populations are subject to different environmental variables. Mutation occurs continually, and although it occurs irrespectively of the needs of the organism, those mutations that confer an advantage in a warm, dry environment tend to be retained by population A, and those that make it possible to leave more offspring in a cool, moist environment tend to be retained by population B. The populations thus diverge genetically, and it may come about that the divergence is reflected in their appearance. The snails of population B, for example, may become darker or larger or differently shaped or more active. Physiological changes, which are not externally visible, may take place. Behavioral differences may manifest themselves: population A may come to prefer the lower surface of vegetation, B the upper surface, and so forth.

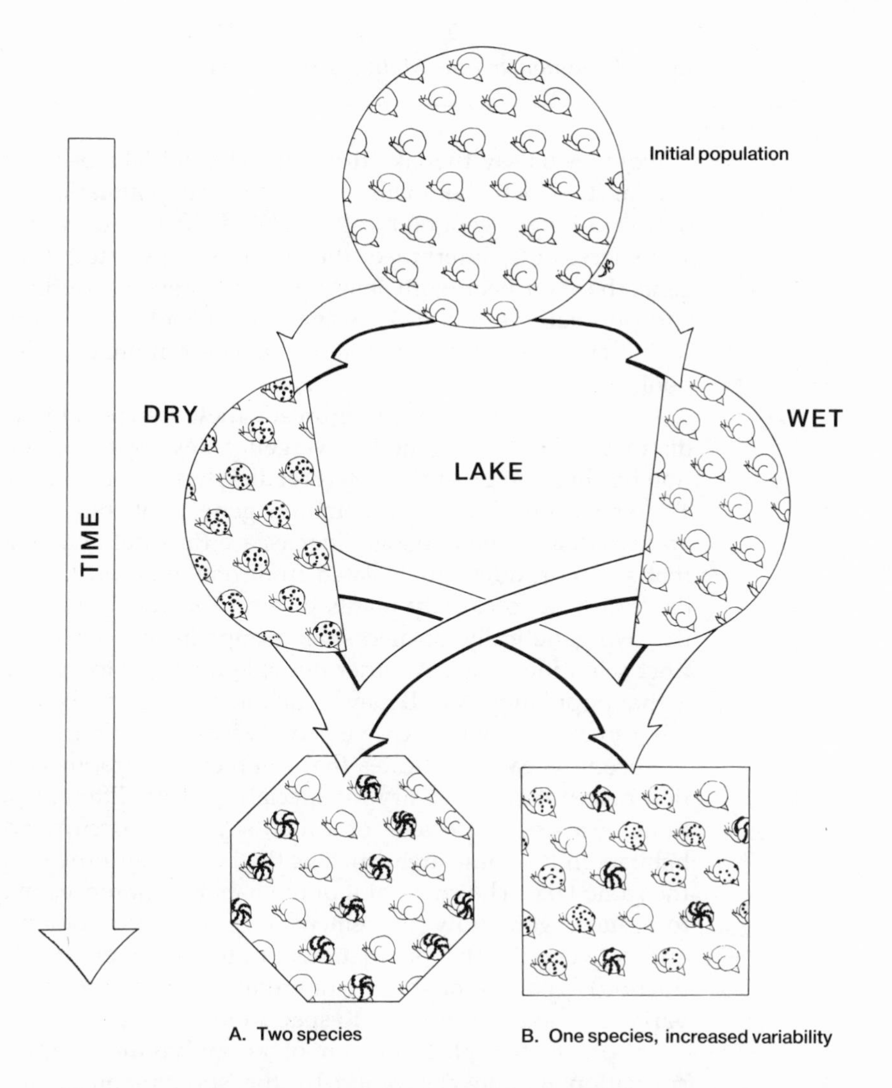

Figure 2. A simplified model of the process of speciation. The model requires two things: genetic variation and geographic isolation. In this example, an initial population of snails is divided by an emerging lake. Genetic variation accumulates in the now isolated populations, which are subject to different environmental conditions (dry and wet) and begin to diverge in character. After many generations, the lake dries up and the two groups gradually resume contact. If enough genetic differences have accumulated, the two forms will be incapable of interbreeding (A), and what had been one species will have become two. If interbreeding does take place (B), only one species can be said to be present, but it has increased its genetic variability beyond that of the initial population.

Now let us say the lake dries up. The old lake bed becomes dry land, which the snails can cross, and populations A and B are no longer geographically isolated. If they come together and successfully interbreed, they have not speciated; the divergence has not been sufficiently great to prevent interbreeding, and populations A and B are again one genetic unit, but a unit with increased genetic variability and a new, more complex gene pool.

But if A and B cannot interbreed, they are considered to be distinct species; the genetic divergence developed under isolation has been so great that interbreeding has become impossible. The snails once shared a common gene pool, but now there are two distinct gene pools. **Species** are thus defined as groups that are reproductively isolated from one another; taxonomists (biologists who classify plants or animals) would give each of the two genetically isolated snail groups in this outcome a distinct scientific name. Another possible outcome to our example is that population A or B may be unable to adapt to the changing environment, in which case it is driven to extinction.

Recent research indicates that complete geographical separation may not be necessary for speciation. Two different populations of a single species—of a fish, say—may occupy adjacent habitats in the same area, such as the bottom and the surface of the same lake. The constraint of the habitat differences may lead to reduced **gene flow** (movement of genetic information within and between the two populations through generations of reproduction), to a lack of sharing of mutations and other sources of variability, and, ultimately, to speciation.

In plants the phenomenon of **allopolyploidy** can lead to speciation in one generation by the spontaneous increase in **chromosome number**. For example, suppose a plant with 14 **chromosomes** hybridizes with a similar species that has 28 chromosomes. A hybrid with 21 chromosomes may be formed (7 + 14, half the chromosomes from each parent). This **hybrid** (a cross between two species) would ordinarily be sterile, because the chromosomes lack pairing partners. (For an explanation of the terms used in this discussion, see Appendix A.) Occasionally, however, due to an accident of **meiosis**, all of the chromosomes enter a **sperm** or **egg** (normally only half do). The resulting **diploid** germ cell can fuse only with another diploid cell. It usually cannot fuse with the normal parental **gamete**, which has only half the chromosome complement (and is said to be **haploid**). If male and female diploid germ cells do come together, a **tetraploid** offspring results. This individual has twice the normal number of chromosomes and is a new species. In the example above, a 42-chromosome species would result. Wheat and other agricultural crops, as well as ornamental plants like tulips and iris, are polyploids; their cells are larger than those of normal plants, owing to their having twice or more the normal amount of genetic material. This often results in a more showy flower or in a greater agricultural potential, which, of course, is what the plant breeder is trying to achieve.

Allopolyploidy is a widespread phenomenon in plants, and the resulting **polyploid** species tend to be more aggressive colonizers of new areas than their diploid ancestors. Polyploidy is not of major importance in the animal kingdom, where, unlike most plants and one-celled organisms, the sexes are separate in the great majority of species.

Reproductive Isolating Mechanisms
Why Are There So Few Hybrids in Nature?

In nature there are a variety of **reproductive isolating mechanisms** that prevent individuals of separate species from hybridizing. In **seasonal isolation**, the annual period of reproductive fertility is different for different species; a fall breeder is not likely to mate with a spring breeder if they come into reproductive condition six months apart. The particular courtship rituals of various animals often yield **behavioral isolation**; the head-bobbing and bill-clapping of a male albatross, for example, do not trigger the mating urge in a female booby. In some closely related groups, the male genitalia may not fit the female; such **mechanical isolation** occurs in many insects. In **gametic isolation** the sperm of the male of one species are not viable in the reproductive tract of the female of another species. And if the sperm of one species does succeed in fertilizing the egg of another species, the developing **zygote** (fertilized egg) may die or be spontaneously aborted owing to gross abnormalities; this isolating mechanism is called **hybrid inviability**. Some hybrids, such as the mule, which results from a mating between a mare (female horse) and a male donkey, are viable, but the offspring (the mule) is sterile. This mechanism, **hybrid sterility**, like all of the other reproductive isolating mechanisms, prevents the exchange of genes between different species—at least beyond the initial hybrid offspring. Natural selection weeds out those individuals carrying the genes that allow the mismatings by ensuring that fertile offspring do not result from such mismating; the "mismating genes" stop there and are not passed on to the third generation. The farmer who wants another mule must once again mate a mare and a donkey.

Occasionally, fertile hybrids do occur. They are not uncommon in fishes. This usually happens under artificial or disturbed conditions and may mean that the two parental species are very closely related, that they are relatively recent derivatives of a common ancestor, or that their isolating mechanisms are incomplete and still evolving. It is not uncommon for most of the fertile hybrids to be of one sex, which further reduces gene flow. Partial isolation, known as **hybrid breakdown**, may result in reduced hybrid fitness, whereby the hybrids, when they mate with one another or with either parent, leave fewer offspring than the non-hybrids do and tend to be eliminated from the population. Many of the viable hybrids that are produced in the laboratories of biologists never occur in nature.

Catastrophes and Mass Extinctions
What Happened to the Dinosaurs?

One of the most exciting and most rapidly developing fields dealing with evolutionary theory concerns **mass extinctions**. Many guesses have been made about the cause of the extinction of the dinosaurs 65 million years ago, at the end of the period paleontologists call the Cretaceous. It has been postulated that mammals ate the dinosaurs' eggs, that the climate became too cold or too hot, that the dinosaurs ate newly evolved poisonous plants, that an epidemic of some sort swept through them, or that some other earthly process did them in. Recent speculation based on a detailed examination of the fossil record of hundreds of marine-animal families over the last 250 million years argues that large-scale extinctions occur about every 26 million years. Astronomers and geologists are in the process of marshaling evidence that shows these cyclic extinctions are caused by periodic showers of comets. The idea is that the impact of a single large comet would cloud the sky with dust, all around the planet, thereby blocking out sunlight and causing the death of a wide range of plant and animal life. The force that drives this system of comet impacts is hypothesized to be an as yet undiscovered, incredibly dense companion star to the Sun that passes close enough to the Sun every 26 million years to disturb the orbits of the comets, causing some to strike the Earth.

A problem associated with the cyclic-comet hypothesis is that the 26-million-year cycle is only an average figure and may be a statistical artifact. All mass extinctions do not neatly fit this cycle, and the Cretaceous extinction of the dinosaurs and many other groups took place not instantly, but over a period of several million years. Different organisms died out at different times, rather than all at once as the comet hypothesis might lead one to expect. Some statisticians doubt the reality of a true periodicity, and some astronomers feel that the hypothesized mechanism is improbable.

A similar scenario involves the impact of a large asteroid on Earth, which may have scattered huge quantities of particulate matter into the atmosphere. Large amounts of iridium, a rare element on Earth, are associated with the Cretaceous-early Tertiary transition, and since iridium is much more common in meteorites than in the Earth's crust, its abundance at the transition is cited as evidence of an extraterrestrial impact that led to the mass extinction of many plant and animal groups, including dinosaurs. Opponents of this hypothesis argue that the iridium was released from deep in the Earth's mantle by a series of intense volcanic eruptions during a relatively short interval of geologic time, and not by the impact of an asteroid. Likewise, data from western North America (Sloan et al., *Science*, 1986, 232: 629–34) indicate that the extinction of the dinosaurs was a gradual process that lasted at least 7 million years, and that the final local extinction occurred about 40,000 years after the postulated asteroid impact. This argues against the comet or asteroid hypothesis as a major cause of extinction.

A reduction of global temperature over the last 15 million years of the Cretaceous and a lowering of sea level combined with massive volcanism are thought to have resulted in an increased seasonality, which would have caused significant deterioration of plant life (Hallam, *Science*, 1987, 238: 1237–42). This shift in food and habitat character, combined with competition from newly evolved mammalian herbivores (plant eaters), is offered as a terrestrial alternative explanation for the extinction of the dinosaurs. (That dinosaurs are extinct is a moot point: they left descendants called birds.)

If one of these interesting suggestions of extraterrestrial agency, such as comet showers or asteroid impact, or the idea of massive volcanism, turns out to be correct, it will have evolutionary implications. It may mean that extinction is not necessarily, or in every case, the result of an inevitable biological process. Extinction may be more a matter of bad timing than of bad genes. But whatever their cause (I think a scenario based on terrestrial causes is more **parsimonious** and therefore more likely than an extraterrestrial one), mass extinctions would provide empty **ecological niches** for surviving species to **radiate** into, thereby promoting evolution, speciation, and diversification.

Embryology, Morphology, and Biogeography
Where Do We See the Evidence for Evolution?

If the reader is still with me, we now have enough background to look for the evidences of evolution in the natural world. Evolution has produced about 2 million living microbial, plant, and animal species that have been described by science, and anywhere from four to thirty times as many remain to be named (if they are not all driven to extinction by the actions of humanity in the meantime). The fundamental unity of this great diversity of life lies in the fact that virtually all organisms carry their genetic information in the DNA molecule, within the cell. *The only reasonable explanation of this fact is that all organisms are related by descent.* There is also a consistency in the chemistry of life beyond the universality of DNA. For example, numerous biochemical compounds perform the same role in **metabolism** in all cells of all organisms. The same sequence of metabolic reactions releases the energy stored in chemical bonds in a great variety of very different organisms. The large molecules used for energy transformations show a remarkable consistency in structure; for example, biochemical pathways such as the Krebs cycle (a stage of cellular respiration), the cytochrome system (proteins necessary for respiration), and others are identical in a wide variety of plants and animals. The same 20 amino acids compose the proteins of all living organisms, and these amino acids are all left-handed molecules. (Organic molecules can exist in two forms that are mirror images of one another; they have the same chemical composition, but one kind rotates a plane of polarized light to the right, the other kind to the left.) Simulation experiments have shown that both right-handed and left-handed amino acids form spontaneously, but only the left-handed molecules have been selected in successful evolution. The universal finding of left-handedness is simply a reflection of the common, historical origin of all life. *This common thread among living things is completely consistent with a theory of descent with modification.*

Comparative **immunology** can be used to show evolutionary relationships. The blood **serum** (the fluid portion of the blood remaining after **coagulation**) of each animal species contains its own set of proteins. If one injects human serum into a rabbit, the rabbit will form **antibodies** (a protein produced in response to a foreign substance) to attack the foreign protein, which is called an **antigen**. Rabbit serum containing these antibodies can then be tested against the serum of other animals. The serum's antibodies will react most strongly to the serum of those animals sharing the most similarities (the same antigens) with the human serum that instigated the production of antibodies in the rabbit; the antibodies will react less strongly where the similarities are fewer. Such antigen-antibody tests show that the nearest living relatives of humans are the great apes, followed by the Old World monkeys, then the New World monkeys. In other words, the great apes share many of the same serum proteins as humans.

Since protein formation is under direct genetic control, many genes are apparently shared by humans and the great apes. In fact, we share with chimpanzees and gorillas about 99 percent of the genes that code for proteins. Other primates share fewer of these genes with us, and if the immunological test had included dogs, chickens, turtles, frogs, and fishes, the reaction would be progressively weaker or even nonexistent. Taxonomists use this technique to show immunological distances and relationships and thereby help to place organisms in a hierarchical arrangement (a tree with limbs, branches, and twigs) that corresponds with the way they evolved through time.

In like manner, an analysis of the alpha chain of the blood's **hemoglobin** shows identical sequences of amino acids in humans and chimpanzees, and a single amino acid difference (out of 141 possible differences) in the gorilla. The relationships inferred from these biochemical and immunological techniques agree very nicely with relationships based on **morphology** (form and structure), which in the past was almost all that taxonomists could draw on in classifying organisms. (Data derived from DNA comparisons among primates will be discussed in Chapter 4.)

Shared similarities and differences are, in fact, the basis for the classification of plants and animals. The reason for the similarities and differences is that some organisms are more closely related to each other by descent than others are. For example, the forelimbs of frogs, crocodiles, birds, bats, horses, whales, and humans show essentially the same bony structures, relationship of parts, and embryological development (Figure 3). They are similar in all these ways because they derive from the same ancestral prototype, which has been modified by natural selection over many millions of years for different functions in different environments. The terrestrial vertebrates (back-boned land animals) are in fact all derived from lobe-finned fishes that had the same arrangement of limb bones as the land animals do (other fishes have rather frail fins incapable of supporting weight out of water).

Other morphological evidence for evolution is presented by

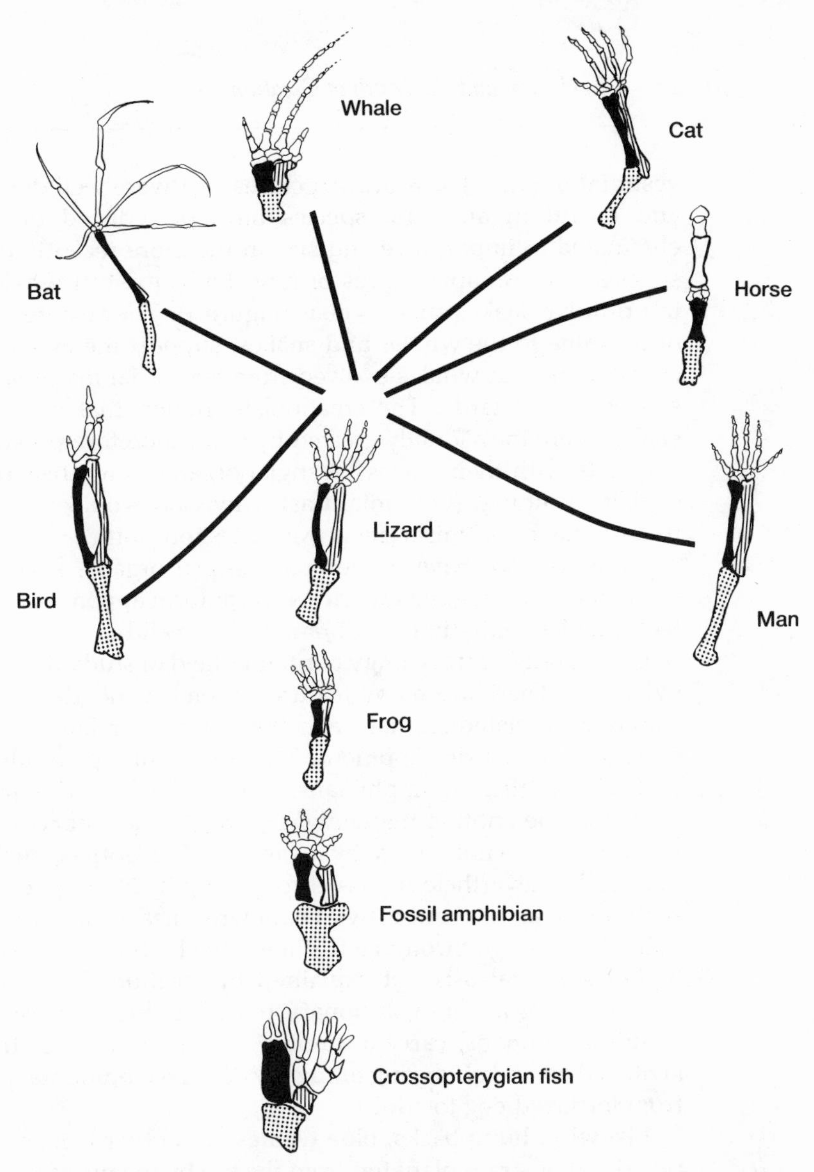

Figure 3. Forelimb homologies of various vertebrates, at different scales. The corresponding bones of the different animals are indicated by shadings and hatchings: all of the stippled areas indicate the humerus bone; the black areas, the radius; hatched, ulna; white, metacarpals (palm) and phalanges (digits). All vertebrate forelimbs, including even those of bats and whales, are structurally similar and are derived from the same embryonic germ layers. The best explanation for this consistent structural plan is that all vertebrates are derived from a common ancestor. (For why the horse has only one digit, see Figures 4 and 12.)

vestigial organs. These are structures that were well developed and useful in ancestral species but are reduced or almost eliminated in importance and size in the more recently derived species. For example, traces of hind limbs exist in whales and in primitive snakes such as boas (Figure 4). The vestiges, surely of no value to the whales and snakes, support the evolutionary explanation that whales evolved from terrestrial mammals, and snakes from lizards. The creationists' notion that whales and snakes were individually created by God, therefore presumably complete with their useless vestigial organs, is not testable and explains nothing. (One might ask why God would want to do that—some creationists say it was to test our faith.)

Humans, too, have numerous vestigial organs, such as tail vertebrae, ear-wiggling muscles, a vermiform appendix, wisdom teeth, and a nictitating membrane (third eyelid).

Comparative **embryology** is another field of study that reflects evolution. There are many features of embryonic development common to related animals, and the closer the relationship, the more similar the development. The early embryos of all vertebrate classes (fishes, amphibians, reptiles, birds, and mammals) resemble one another markedly (Figure 5). The embryos of vertebrates that do not respire by means of gills (reptiles, birds, and mammals) nevertheless pass through a gill-slit stage complete with aortic arches and a two-chambered heart, like those of a fish. The passage through a fishlike stage by the embryos of the higher vertebrates is not explained by creation, but is readily accounted for as an evolutionary relic. The higher vertebrates, including humans, carry a number of ancestral genes that are switched on and off during **ontogeny** (the developmental process from fertilized egg to adult).

Likewise, humpbacks, blue whales, and other baleen whales (whales that strain plankton from the sea by means of fine strips of whalebone) lack teeth as adults. Nevertheless, their embryos possess rudiments of teeth. This is understandable only if the baleen whales evolved from the toothed whales (as they surely did).

The fact that birds evolved from reptiles is reflected by a recent discovery that reptile-like teeth can be grown from chick embryos. Certain early fossils, intermediate between reptiles and

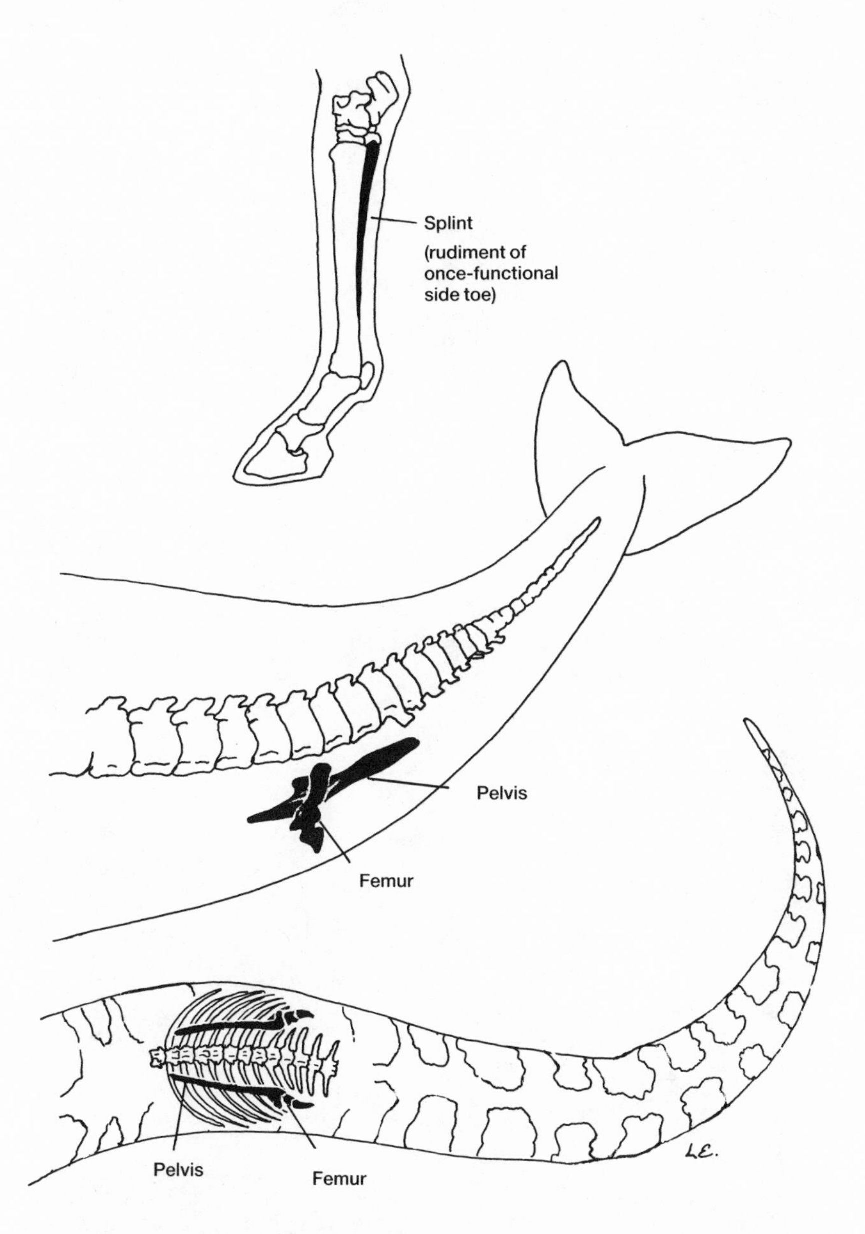

Figure 4. Vestigial structures reflecting common evolutionary history. Pythons (bottom) and whales (middle) have useless rudimentary pelvic girdles and femurs (thigh bones), because they are derived from four-legged ancestors. Modern horses (top) have one toe in contact with the ground and two side toes fused to the leg as splints. This structure reflects descent from three-toed ancestors. See also Figure 12.

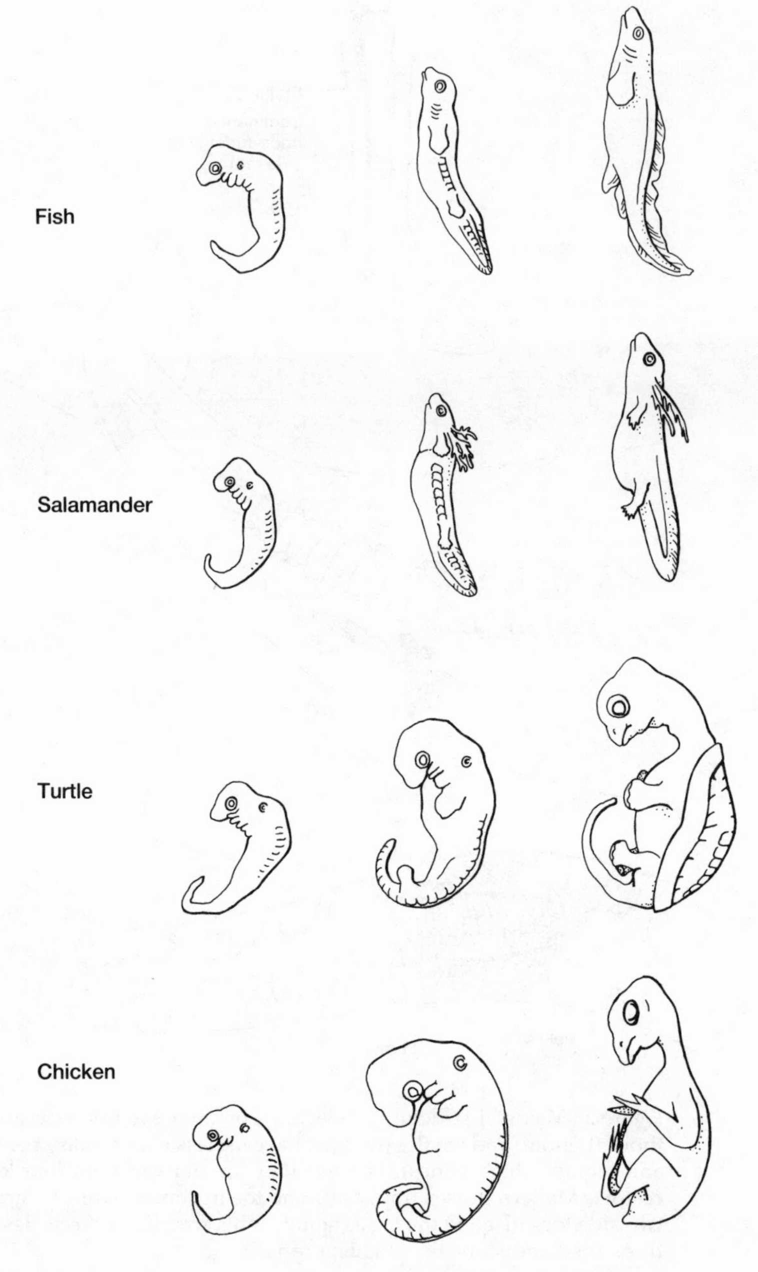
1
2
3
Fish
Salamander
Turtle
Chicken

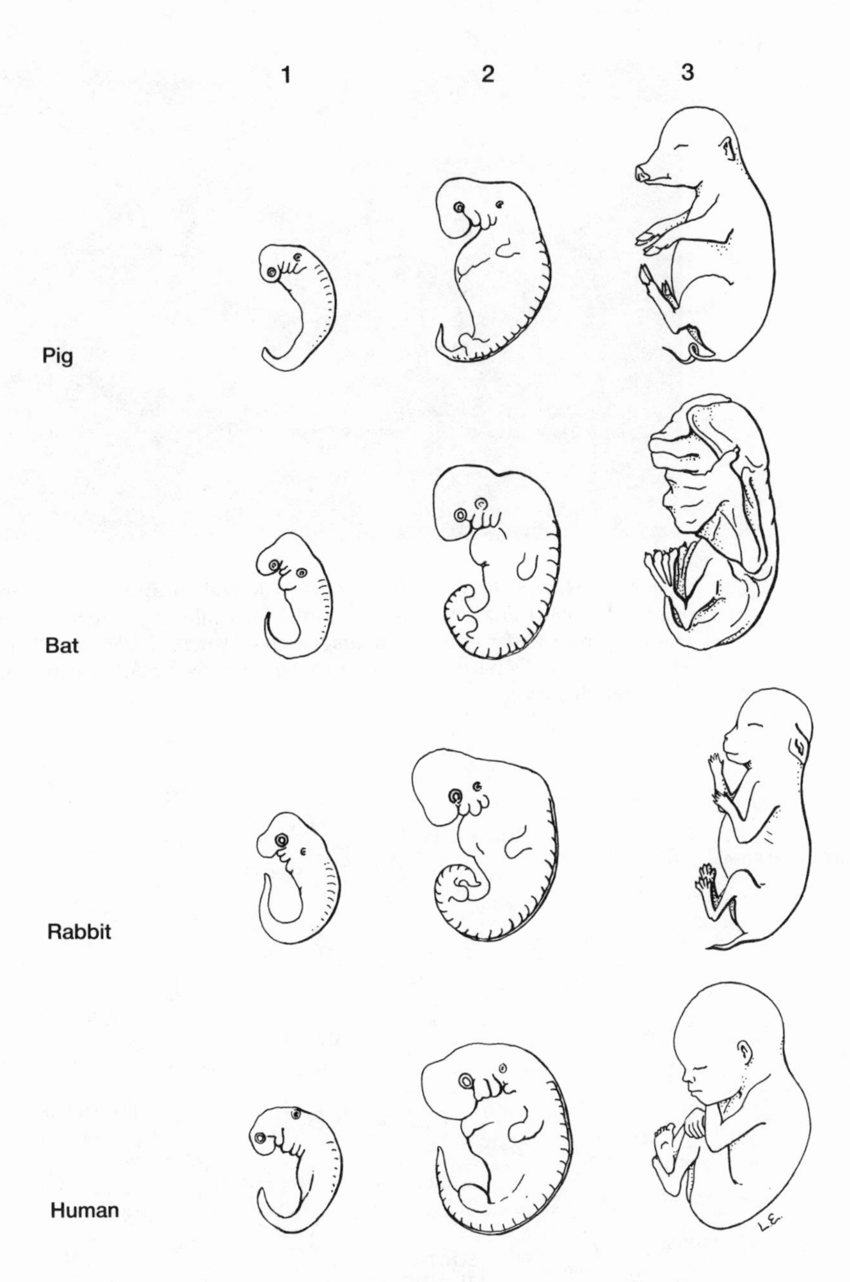

Figure 5. Three parallel stages in the development of eight vertebrates. Note how consistently the early embryonic stages resemble one another, a reflection of common ancestry and descent with modification. (Bat, rabbit, and human drawings modified by permission from Putnam Publishing Group.)

Figure 6. H.M.S. *Beagle* in the Straits of Magellan under the command of Captain Robert FitzRoy. Mount Sarmiento, Chile, 7,500 feet, is in the distance. (From *Order of the Proceedings at the Darwin Celebration*, Cambridge, 1909; courtesy Department of Library Services of the American Museum of Natural History.)

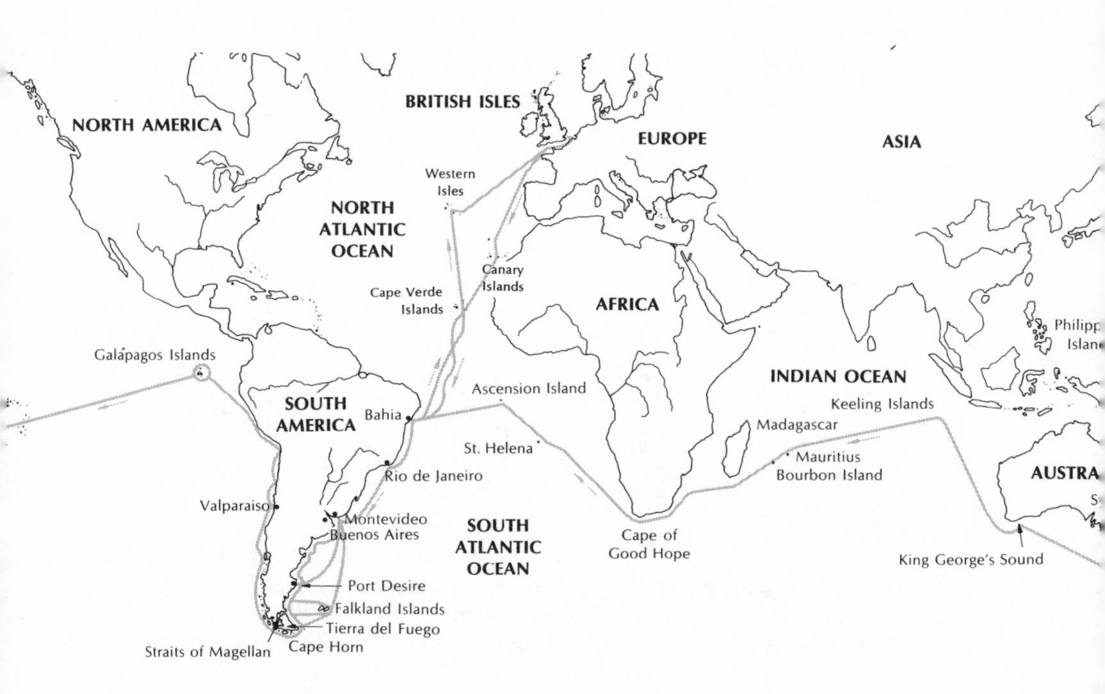

birds, also show the presence of teeth. Thus, some of the genes necessary for tooth development have been retained by some modern birds, but are not normally expressed.

The sciences of embryology and paleontology have combined to reveal that the malleus and incus (hammer and anvil of the mammalian middle ear) evolved from the articular and quadrate bones of the reptilian jaw joint. This transformation of jaw bones to ear bones can be traced in the fossil record via the mammal-like reptiles.

Biogeography, or the study of the geographical distribution of organisms around the Earth, also reflects descent with modification. Darwin called attention to the fact that the plants and animals of remote volcanic islands resemble the flora and fauna of the nearest land mass. He noticed this resemblance while traveling around the world as the unpaid naturalist on board H.M.S. *Beagle* from 1831 to 1836 (Figures 6 and 7). Darwin was astonished by the organisms of the Galapagos Archipelago, a group of 16 islands 600 miles west of Ecuador, right on the equator. The *Beagle* spent five weeks in the Galapagos in 1835, and Darwin visited four islands. Although he did not fully understand all of the evolutionary implications of what he saw there until nearly a year and a half later, the fauna of the Galapagos Islands provided him with some strong hints about evolution.

What Darwin observed was that the Galapagos finches, tortoises, iguanas, and other animals are distinctive, yet bear an unmistakable resemblance to South American forms. He was

Figure 7 (left). Map of the voyage of H.M.S. *Beagle*, 1831–36. Charles Darwin was the 23-year-old unpaid naturalist aboard the *Beagle* when she began her five-year voyage of discovery around the world. This great adventure afforded Darwin the opportunity to collect fossils and to explore coral reefs, deserts, tropical rain forests, the pampas, the high Andes, and many other biomes in a variety of biogeographical realms. His observations and research during this time led to the publication of his popular *Journal of Researches (Voyage of the Beagle)* in 1839, and spawned the development of his revolutionary theory of evolution, published in *On the Origin of Species* in 1859. (Reproduced from C. P. Hickman, Jr., L. S. Roberts, and F. M. Hickman, 8th ed., 1988, *Integrated Principles of Zoology*, by permission of C. V. Mosby Co., St. Louis, as modified from A. Moorhead, 1969, *Darwin and the Beagle*, Harper & Row, New York.)

Figure 8 (facing page). Darwin's finches, the classic example of adaptive radiation, or the evolutionary diversification of a single lineage into a variety of species. These small, drab birds of the subfamily Geospizinae on the remote Galapagos Islands, 600 miles west of Ecuador, initially puzzled Darwin with their overall similarity but highly variable bill size and shape. Later reflection on these variations, and on his impressions of the entire flora and fauna of the Islands, were part of his inspiration for the theory of evolution by means of natural selection. In *Voyage of the Beagle* Darwin wrote, "The most curious fact is the perfect gradation in the size of the beaks of the different species of Geospiza, ... Seeing this gradation and diversity of structure in one small, intimately related group of birds, one might really fancy that, from an original paucity of birds in this archipelago, one species had been taken and modified for different ends."

The finches are (1) woodpecker finch (*Cactospiza pallida*), (2) mangrove finch (*Cactospiza heliobates*), (3) large tree finch (*Camarhynchus psittacula*), (4) medium tree finch (*Camarhynchus pauper*), (5) small tree finch (*Camarhynchus parvulus*), (6) vegetarian finch (*Platyspiza crassirostris*), (7) warbler finch (*Certhidea olivacea*), (8) Cocos finch (*Pinaroloxias inornata*), (9) large ground finch (*Geospiza magnirostris*), (10) medium ground finch (*G. fortis*), (11) small ground finch (*G. fuliginosa*), (12) sharp-beaked ground finch (*G. difficilis*), (13) large cactus ground finch (*G. conirostris*), and (14) cactus ground finch (*G. scandens*). As reflects their habits, the 14 can be divided into tree and ground finches, the latter closest to the ancestral type.

The various species are distinguished by the size and shape of their bills and by body size, but these distinctions are not always easily made, owing to the similarity among species and the variation within species. Whereas some species are widespread throughout the Islands (*G. fuliginosa* breeds on 14 of the 17 major islands), others have a more restricted range (*Camarhynchus pauper* occurs on only one island). Some islands support breeding populations of ten species of finches, while others, the smaller islands, are home to only three. No species of Darwin's finches has yet become extinct, but some local populations have already been eliminated by human activity.

Bill size and shape are related to diet. Big-billed finches can eat larger food items; the ground finches' bills, deep at the base, are adapted for crushing hard seeds; the curvature of the bills of the tree finches is useful for picking arthropods from their hiding places; the bills of the warbler and woodpecker finches are ideal for probing flowers and woody holes.

Some of Darwin's finches show amazing adaptations. The woodpecker finch uses cactus spines to dig insect larvae out of dead branches, one of the rare examples of tool use by a non-primate. *G. difficilis* pecks wounds on boobies and other large seabirds and drinks their blood like a vampire. Other *Geospiza* species reverse that role by picking parasites off the skin of Galapagos iguanas and tortoises.

This diversity of form and function arose relatively recently. The Islands, of volcanic origin, emerged not more than 5 million years ago. Study of protein variation in the finches suggests that the adaptive radiation of the 14 species from a common South American ancestor occurred just one-half to one million years ago. For further details, consult Grant (1986), under Further Reading. (Drawing by Eric Mose, from "Darwin's Finches," by David Lack. Copyright © by Scientific American, Inc. All rights reserved.)

1
2
3
4
5
6
7
8
9
10
11
12
13
14

struck by the fact that the shells of the giant tortoises, for which the Galapagos Islands are named, varied from island to island. After he returned to England and had a chance to examine and think about the specimens of finches collected from the Galapagos (Figure 8) he realized that they too were different from island to island, especially with respect to the size and shape of the beak, yet they were otherwise quite similar to one another and to mainland finches. He eventually concluded—correctly—that a small number of ancestral finches from the mainland, blown far to the west by storms, had colonized the newly emergent Galapagos Islands. Because they encountered no competition, the finches radiated into a variety of functional roles in the community; each of these roles represents a concept known as an **ecological niche**, each niche presenting its own environmental pressures. The formation of new species from the colonizing ancestors followed. Much the same thing must have happened when the tortoises and iguanas arrived, probably on rafts of floating vegetation carried west from South America on ocean currents.

Darwin recalled that the organisms of the Cape Verde Islands off Africa, which he had observed earlier, resembled African forms, even though the environmental conditions were very similar to those of the Galapagos. He had begun his voyage with the traditional idea that every species was created in place, but if special creation were really how things came to be, there would be no reason for species on volcanic islands to resemble the inhabitants of the nearest land mass. Darwin's insight into the **mutability** of species—his insight into their potential for change —*was the beginning of the end for the idea that species were individually created and forever immutable*.

If one is to deal with the theory of evolution rationally, the above-mentioned evidences for evolution from biochemistry, immunology, morphology, embryology, paleontology, and biogeography cannot be ignored or explained away. *They all point to the fact that evolution has occurred*. The fossil record, which presents some of the most dramatic evidence for evolution, is the subject of the next chapter.

2 Geologic Time and the Fossil Record

Fossils are the remains or traces of prehistoric life. There are fossils of species that still survive today, but most fossils are of animal or plant species long since extinct (see Figure 9). By studying great numbers and varieties of fossils, we can see evolution in action over far greater time periods than would be possible if we were to restrict ourselves to living groups of plants and animals. *Fossils provide hard evidence that evolution has occurred.*

The hard parts of animals, such as teeth, bones, and shells, are the most likely to become fossilized, but there are also some fossils of soft-bodied organisms such as jellyfish, as well as fossilized imprints of leaves and other plant parts. There are complete insects, lacking only the soft interior parts, preserved in **amber** (a hard, translucent, fossilized plant resin). There are even fossilized footprints and feces: the latter are called **coprolites**. In fact, fossil feces can yield a great deal of information, including the diet of an extinct animal. The habitat and plant associations in which an animal lived can be deduced from the seeds found in coprolites, and deductions of this sort can help geologists and other scientists reconstruct ancient landscapes and climatic conditions.

But the chances of something's becoming fossilized are not great. The remains of an animal or plant must first escape destruction by scavengers and physical forces and then become buried in **sediment**, such as at the bottom of a lake or in volcanic

Figure 9. A fossilized specimen of the herring-like fish *Diplomystus dentatus* from the Green River Formation of the Wyoming Eocene. This specimen (the fish is just 3 inches long) shows the sort of delicate detail that is often revealed in fossils; and its location demonstrates what vast changes have taken place in North America in 50 million years. (Courtesy of Lance Grande and the Geological Survey of Wyoming.)

ash. This enveloping material must then survive the routine folding, crushing, and erosion that occurs in the rock formations across the millions of years. Finally, the rock in which the fossil is preserved must be exposed and come to the attention of a **paleontologist** (a scientist who studies prehistoric life). The fossil record is therefore understandably incomplete; reading it is like reading a book that is missing many of its pages and even some whole chapters—with some hard work and thoughtful reflection, we discover that the pages and chapters we do have provide a very useful window from which to view the past. We know the record is incomplete, but from fragments of solid evidence we know that certain conclusions can safely be drawn, and as new pages and chapters are discovered, we steadily improve the picture of the past—much in the manner of a detective reconstructing a crime.

When Charles Darwin published *On the Origin of Species* in 1859, the fossil record was already well enough understood that scientists recognized the "progressive" nature of a very old Earth. It was clear that some things lay atop others, that some were therefore presumably older than others. Evolutionary theory supports that observation, predicting that when the remains of plants or animals are present in rocks of different ages, then these remains should describe some sensible sequence reflecting the extinction of some groups and the evolution of new groups as time passes and the environment changes.

Superposition and Faunal Succession
What Do the Layers of Rock Tell Us?

If you look at a wall of the Grand Canyon or even the face of a roadcut through a mountainside, right away you will notice variously colored layers of different materials. These are rock **strata** that were deposited at different times. In undeformed **sedimentary rock** the oldest strata were deposited first and are therefore at the bottom, and the youngest rocks are at the top. This is called **superposition** and provides a relative age for each stratum. The record of the rocks piled up in a particular location is called a **stratigraphic sequence**, and the scientist who studies these sequences is called a **stratigrapher**. This system, called **stratigraphy**, was used well before Darwin's time and does not depend on the fossil record for its validity. Geologists did not then know the actual age of any of the strata, but they knew which ones were older; they also knew that fossils would add their own kinds of information to their efforts to reconstruct the past, but many of the rock strata they examined contained no fossils.

The early stratigraphers noticed that when fossils of particular organisms were found, they were embedded in particular strata; fossils of other organisms were associated with other strata. In general the fossils show a succession from **primitive** forms in the older (lower) rocks to more **advanced** forms in the younger (higher) rock. Thus in very old strata there are many fishes and no mammals. But modern paleontologists look also to the structure, or **morphology**, of the forms, not simply their stratigraphic position, or age, in concluding that one form is primitive, another more advanced. (Primitive does not necessarily mean extinct; many primitive animals and plants, such as scorpions and ferns, persist today.)

Particular fossils that keep turning up in different parts of the same stratum, perhaps hundreds or thousands of miles apart, are called **indicator fossils**, because they help geologists recognize the equivalent age of rock formations in separate parts of the world, and thus provide an independent means of arriving at a relative age for each stratum. Though this procedure cannot establish the **absolute age** of the rock strata (more on that later), it succeeds very impressively in demonstrating that Rock Stratum A is older than Rock Stratum B.

The principle that fossils in a stratigraphic sequence succeed one another in a definite, recognizable order is called **faunal succession**. (If the fossils are of plants, the sequence is called a **floral succession**.) By using superposition and faunal or floral succession, geologists and paleontologists constructed a geologic time table (Table 1) for many regions of the world, a timetable that was essentially established by 1841, 18 years before Darwin published *On the Origin of Species* (see Appendix B). Thus the ground-

TABLE 1
Geologic Time

Eras, Periods, and Epochs	Span of time covered (in millions of years before present) From	To	Emerging groups, or events
Cenozoic			
Quaternary			
Recent	0.01 MYA	today	Retreat of last ice age (epoch covers just last 10,000 years)
Pleistocene	2.0	0.01	*Homo* (modern human)
Tertiary			
Pliocene	5.1	2.0	*Australopithecus* (ape-man)
Miocene	24.6	5.1	Apes
Oligocene	38	24.6	Monkeys
Eocene	54.9	38	Lemurs, lorises, tarsiers
Paleocene	65	54.9	Origin of primates
Mesozoic			
Cretaceous	144	65	First flowering plants (period closes with extinction of dinosaurs)
Jurassic	213	144	First birds; abundant dinosaurs
Triassic	248	213	First dinosaurs, mammal-like reptiles; origin of mammals
Paleozoic			
Permian	286	248	Radiation of reptiles
Carboniferous	360	286	First reptiles
Devonian	408	360	First amphibians
Silurian	438	408	First jawed fishes
Ordovician	505	438	Spread of jawless fishes
Cambrian	590	505	Trilobites and other hard-bodied invertebrates
Precambrian			
Ediacaran	4,600	590	First multicellular fossils of jellyfish, sea pens, wormlike animals, and algae

SOURCE: Revised geologic time scale based on Harland et al., 1982.

work for a theory of change in the living things of the Earth was laid long before Darwin began piecing together a formal theory of evolution.

Radioactive Decay and Absolute Dating
How Accurate Are Natural Clocks?

Until well into this century, geologists and paleontologists had to be content with a knowledge of the *relative* age of a rock stratum; it was older than this one, younger than that one, and the fossils it contained allowed scientists to relate its age to other rock in other parts of the world. But how old it might really be, or the Earth itself might be, could not be determined by any scientific means then available. Today, however, we are able to determine the *absolute* age of ancient rocks, whether they contain fossils or not, by means of a technique called **radiometric dating**. All rocks are composed of minerals, and many minerals contain a radioactive isotope (**radioisotope**) of an element such as uranium that acts as a natural clock. One by one, at an absolutely predictable rate, the radioactive atoms of a radioisotope spontaneously decay (break down) into atoms of a new, nonradioactive material. Because of this constant decay, the radioisotope is said to be **unstable**; it changes over time. What it changes to, called the **disintegration product**, is a different element, stable forever in its composition. The decay of the radioisotope continues over vast periods of time, but at a fixed rate, independent of temperature, pressure, or other environmental variables. *No other process on Earth proceeds at such an utterly constant rate.*

The decay rate that is so dependable in the absolute dating of rocks is expressed as the **half-life** of the radioisotope; that is the time it takes for half of the radioactive atoms in the sample to decay. For example, let's say that unstable isotope A decays to stable isotope B with a half-life of 1,000,000 years. If we start with 1,000 atoms of A, then at the end of 1,000,000 years, we will have 500 atoms of A and 500 atoms of B. At the end of 2,000,000 years, half of the remaining atoms of A would have decayed, and we would be left with 250 atoms of A and 750 atoms of B, and so on. If we are given a sample of a rock, and we find that it contains 125 atoms of A and 875 atoms of B, then we can determine quite directly that the rock was formed 3,000,000 years ago. All we need is some means of accurately counting the atoms of A and the atoms of B, for if we know the half-life of a radioisotope, and can measure the amount of the radioactive element remaining and the amount of the disintegration product present, we can back-calculate to the time the sample was originally formed, when none of it had begun to decay.

Today, science has such a means. Several radioactive isotopes are used in determining geologic time. A **mass number** (the sum of the protons and neutrons in the **nucleus** of an atom) designates each isotope. Thus we have uranium 238, which decays to lead 206 with a half-life of 4.5 billion years. Uranium 235 decays to lead 207 with a half-life of 704 million years. Potassium 40 decays to argon 40 with a half-life of 1.25 billion years. The potassium-argon system is relatively common in rocks and is therefore more widely used than other systems. Each of these decay processes (there are several others, as well) offers us a system of dating, and in each case the extent of the decay can be measured in the laboratory with great precision. The use of more than one of these systems on a given rock sample helps ensure the accuracy of the dating. (Still, if knowing the rock's age is to be of any value, it is crucial that the geologist has recorded precisely where the rock sample was taken from the ground.)

The theory of radiometric dating is very simple, but in practice the technique is complicated by difficulties in making precise measurements of tiny amounts of isotopes. Moreover, not all rocks can be used for radiometric dating. **Igneous rocks** (rocks that cooled from the molten state), such as volcanic or granitic rocks, are the best for radiometric dating. Fossils do not occur in such rocks, but if fossils occur *between* two undisturbed layers of volcanic rock, the age of the fossils has to be between the dates of the two volcanic strata. Some skeletons, if they have incorporated sufficient amounts of uranium, may be dated directly. Measurements of the decay of uranium 235 to lead 207, even in very small samples of fossils, are accurate to within 2 percent, but 2 percent of a billion years is a lot of time, and a corroboration from stratigraphy can be valuable.

Radiocarbon (carbon 14) dating can be applied directly to fossils or organic human artifacts (but not stone artifacts) and can be quite accurate. Unstable carbon 14 is formed in the atmosphere from stable nitrogen 14 by the bombardment of cosmic rays from space. Because its rate of formation balances its rate of decay, in atmospheric carbon dioxide the proportion of unstable carbon (^{14}C) relative to ordinary carbon (^{12}C) is essentially constant. Plants take in the atmospheric carbon dioxide from the air, and animals eat the plants. Thus both plants and animals have a fixed amount of ^{14}C in their tissues while alive, but after death no new ^{14}C can replace the amount lost by radioactive decay to ^{14}N. Therefore, by measuring the remaining proportion of ^{14}C relative to ^{12}C, we can calculate the approximate time of death as far back as about 50,000 years ago. In a sample older than that, the amount of ^{14}C remaining today is too small to be accurately measured (the half-life of ^{14}C is just 5,730 years). Creationists attempt to discredit ^{14}C dating by applying it to fossils older than 50,000 years, or in other inappropriate ways, and then showing that it yields obviously erroneous dates.

Reading the Rocks and the Fossils

How Old Is the Earth, and How Old Is Life?

By combining the results of the absolute dating method (calculating radioactive decay) with those of the relative dating method (seeing what rock strata lie above and below), we can interpret the fossil record with increasing confidence. *The age of the Earth and Moon has been demonstrated to be about 4.5 billion years.* This is not a guess based upon a few selected rocks. Over the last 30 years nearly 100 independent laboratories worldwide have published in the scientific literature over 100,000 radiometric ages that support this conclusion.

The earliest fossils discovered to date are bacteria and primitive plants called blue-green algae, which date back to about 3.5 billion years. Other Precambrian fossils (see Table 1) include sponge spicules and impressions of jellyfish, soft corals, and segmented worms. We are lucky to have any fossils this old, because rocks of such antiquity have been greatly folded and distorted, and because the earliest animals were mostly soft-bodied, with few hard parts to fossilize. Cambrian rocks, however, from 570 million years ago (MYA), are rich with the remains of many invertebrate groups, such as protozoans, sponges, jellyfish, and various shelled organisms.

As one moves forward in geologic time, up into higher rock strata, other groups of animals are encountered. The first known backboned animals, primitive fishes called ostracoderms, appeared in the late Cambrian, over 500 MYA, the first jawed fishes (placoderms) in the Silurian, the first amphibians in the Devonian, the first reptiles in the Carboniferous. The Mesozoic is known as the "Age of Reptiles" for its proliferation of dinosaurs and other reptiles. The first known mammals and birds split away from different reptilian ancestors in the Triassic and Jurassic, respectively. Primates, the mammalian order to which we belong, arose in the Paleocene and were widespread by the Oligocene, about 38 MYA.

This sequential appearance of different groups at different times, the more advanced appearing in general later than the more primitive, is predicted by evolutionary theory. *It cannot be reconciled with creationism,* which requires all groups to have come into existence essentially simultaneously and fully formed a mere 6,000 to 10,000 years ago. People once thought that the Earth was flat and that it was the center of the universe. It is now high time we laid the creationist ideas to rest, as well.

Transitional Fossils

Are There Really Missing Links?

Hundreds of books have been written about fossils, dozens more appear every year, and we cannot expect to cover all facets of paleontology in a short essay. Two examples, however, will go far toward illustrating the kinds of information that can be obtained from fossils. These examples (and there are many others) demonstrate that fossils intermediate between major groups do exist, as predicted by evolution. Creationists deny that such things can exist, but the evidence is in the rocks, in countless numbers of fossils. Zoologists may classify these intermediate animals as belonging to this group or that group, but they do so only because they must draw the line somewhere, not because they doubt their transitional nature. The line they draw is a conceptual convenience, not a problem for the theory of evolution.

The most famous **transitional fossil** is *Archaeopteryx* (Figures 10 and 11), a crow-sized animal that dates back to the Jurassic about 150 MYA, and is known from skeletons retrieved from the limestone of Bavaria. Today, *Archaeopteryx* is classified as a bird, because the fossils clearly show the impression of feathers, but before the feather impressions were noticed, it was classified as a reptile on the basis of its skeletal structure. What kind of reptile? That is a question much debated by paleontologists. The most likely ancestors of *Archaeopteryx* were either the theropods, small dinosaurs that walked on their hind legs, or the pseudosuchians, an ancient group that also gave rise to the crocodilians. Birds are in fact sometimes referred to by paleontologists as "glorified reptiles," since *Archaeopteryx* had such reptilian (and unbirdlike) characteristics as toothed jaws, clawed fingers (three on the front of each wing), abdominal ribs, and an elongated bony tail. In addition to feathers, its birdlike features are a furcula (wishbone) and a bird's pelvis. *Archaeopteryx* is clearly intermediate between reptiles and birds. It may not itself be the ancestor of modern birds, and it may not be precisely midway between reptiles and birds, but it demonstrates the transition—clearly it or something like it, descending from reptiles, was the forerunner of birds.

Did it fly, glide, or run? This is another puzzle for paleontologists. There is some evidence that flying, or at least gliding, was possible. *Archaeopteryx's* forelimb feathers are asymmetrical, like those of all flying birds. (Non-fliers, like the ostrich, have symmetrical forelimb feathers; the surface area on the right side of a non-flier's feather shaft is the same as the surface area on the left side.) On the other hand, *Archaeopteryx* lacks some of the structural adaptations of flying birds, such as a sternum to support flight muscles. It did, however, have a very large furcula, on which the chest muscles (those that birds use for flight) originate. Feathers evolved from reptilian scales, and their original virtue may have been to insulate the body. Whatever their original function, their development eventually permitted flight. Most likely, a running or jumping stage preceded and led, in time, to short glides and, ultimately, to the development of true flight.

So there is much we do not know about *Archaeopteryx*, but there is also much we *do* know, *and not to perceive its transitional nature is to be willfully blind to the obvious*.

Archaeopteryx has recently been subjected to a careful reexamination because of allegations by British astronomer Fred Hoyle and mathematician Chandra Wickramasinghe that the fossil's feather impressions are a forgery. Hoyle and Wickramasinghe, it should be noted, have no expertise in biology or paleontology, do not understand natural selection, and have espoused a variety of anti-evolutionary ideas. Because these allegations, regardless of their lack of merit, could give aid and comfort to the creationists, scientists at the British Museum (Natural History) subjected *Archaeopteryx* to a battery of tests that confirmed its authenticity beyond doubt. In one such test, hairline cracks on feathered areas of opposing slabs of the fossil were exactly matched. This rules out Hoyle's claim that modern feathers were pressed into an artificial cement layer. No such layer was present. The complete tests are described in an article by Charig et al. in *Science* (1986, 232: 622–25). In November 1987 a new specimen of *Archaeopteryx*, the sixth and largest known specimen, was discovered in a private collection in Solnhofen, West Germany (see Wellinhofer in *Science*, 1988, 240: 1790–92). In the Solnhofen fossil are imprints of feather shafts, further supporting the fact that the feather impressions on the other specimens are not forgeries, as Hoyle and Wickramasinghe had alleged. Hoyle's motives are unclear, but he would not for a moment brook a comparable challenge in his own area of expertise. His claim demonstrates what can happen when a scientist makes uninformed remarks about a subject in which he is unqualified.

Our second fossil example is the succession of animals making up the horse family, called by zoologists Equidae. This group provides the most complete series of fossils in an animal lineage. The scene opens in North America in the Eocene about 54 MYA,

Figure 10. Archaeopteryx lithographica, the 150-million-year-old ancestor of modern birds, from the Jurassic limestone of Bavaria. The impressions made by the feathers are quite apparent. The skeleton alone, if we ignore the feather impressions, is largely reptilian, indicating that this animal is a transitional form between reptiles and birds. (Museum für Naturkunde, East Berlin, German Democratic Republic)

Figure 11. Artist's reconstruction of *Archaeopteryx*, based on the details preserved in the fossil (see Figure 10). (Painting by Rudolf Freund, courtesy of the Carnegie Museum of Natural History, Pittsburgh, Pa.)

with a tiny, terrier-sized, forest-dwelling browser called *Hyracotherium* (Figure 12). It had four functional toes on the front feet and three on the hind feet, and the teeth were short-crowned, indicating that it ate relatively tender foods. As the tropical vegetation of central North America gradually gave way to open, grassy plains, size and speed became increasingly important, if the lineage was to avoid extinction at the jaws of predators. *Mesohippus* was very common in the Oligocene; it had three toes on each foot, and its teeth show that it too was a browser, but it was bigger than *Hyracotherium*.

The Miocene period saw a proliferation of horse species as grassy prairies became widespread under the influence of a drier and cooler climate. (The tree of horse evolution actually is better described as a bush, because of its many branches; the branches mentioned below led to modern horses, but there were many other branches that became extinct.) *Merychippus* was a Miocene horse, about the size of a pony, that had high-crowned teeth with complicated enamel ridges, indicating that it was a grazer on tough grasses rather than a browser of tender vegetation. It had three toes on each foot, but only the middle digit touched the ground. The elongation of toes and the reduction in number of toes constitute an adaptation for speed in an open environment. Longer limbs result in a greater stride length and therefore greater speed. With *Pliohippus*, which succeeded *Merychippus*, toe reduction and elongation were carried further: the two side toes on each foot were reduced to mere splints.

Modern horses, *Equus*, larger and no doubt stronger and faster than *Pliohippus*, demonstrated the effectiveness of their adaptation by dispersing during the Pleistocene from North America to South America, Eurasia, and Africa. Near the end of the Pleistocene, horses became extinct in North and South America. We do not know the exact reasons because it is difficult to choose from among several possible explanations, such as changing climate and human predation. The horses and donkeys we have in the Americas today are in fact the descendants of stock reintroduced from Europe by the Spanish explorers in the sixteenth century. In Africa the horse family survives wild as zebras. Horses, zebras, and donkeys, though different species, are far more closely related to each other than to any of their distant ancestors.

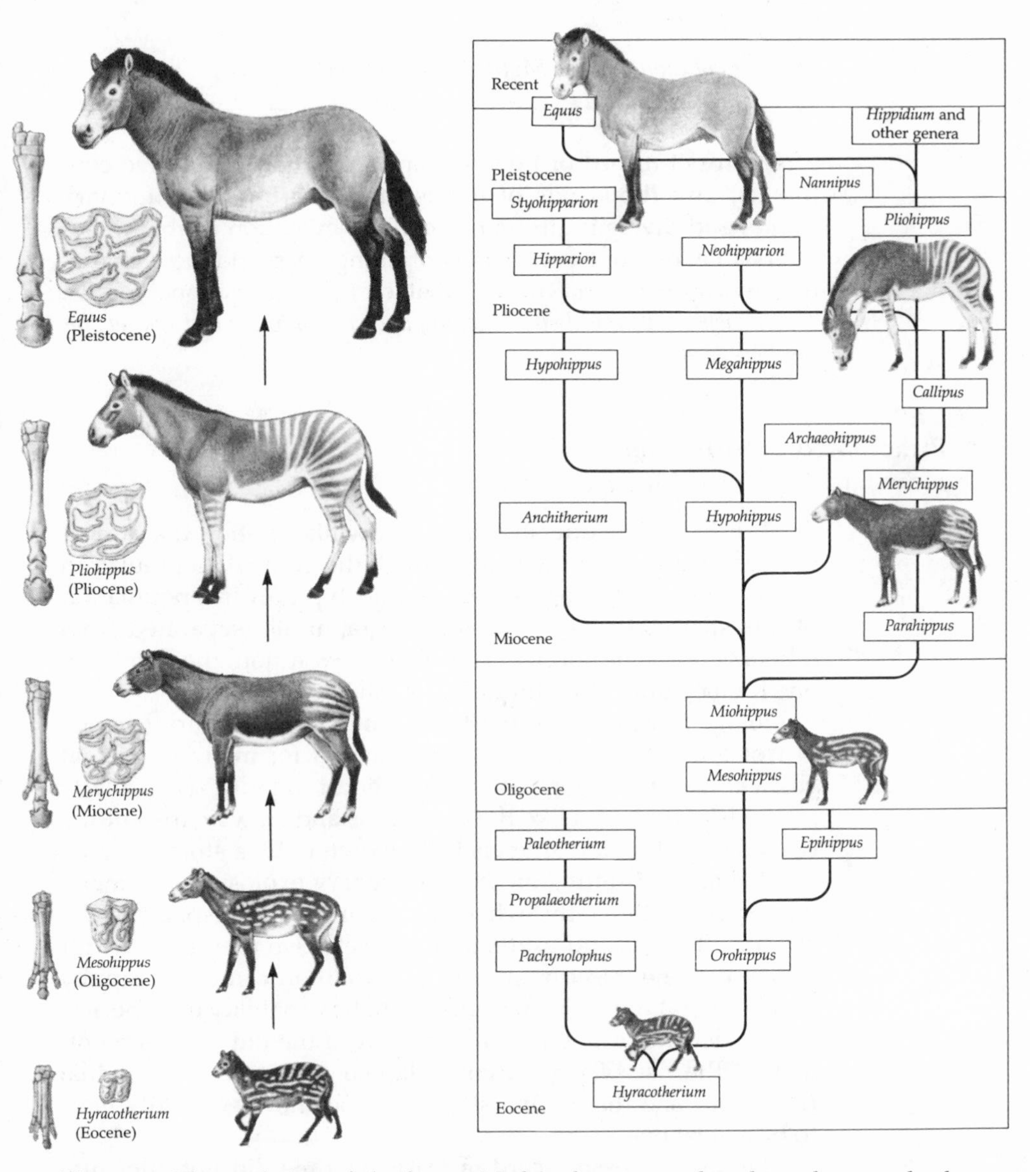

Figure 12. The evolution of the horse. Modern horses are thought to have evolved through stages similar to those depicted at the left, from the small, multi-toed, forest-dwelling *Hyracotherium,* with simple enamel ridges on teeth suitable for browsing tender vegetation, through medium-sized forms with lateral toes that do not touch the ground, to large modern, grassland-dwelling *Equus,* with one toe and complex enamel ridges adapted for chewing tough grasses. The more detailed phylogeny at the right shows that the modern horse and its close relatives, like the zebra, are the only surviving members of an old and complex evolutionary bush with multiple divergent branches. (Drawing from N. A. Campbell, 1987, *Biology,* by permission of Benjamin/Cummings Publ. Co., Menlo Park, Calif.)

The fossil record of horses spans 54 million years, five continents, and thousands of fossils. It shows both evolutionary trends and diversity through time. The evolution of the horses offers an excellent example of continuing adaptation to a changing environment, an evolution that left many transitional fossils. *This lavish fossil record speaks loudly and clearly to the fact that evolution has occurred.*

Punctuated Equilibrium
How Fast Can New Lineages Evolve?

Most biologists operate on the Darwinian notion that a new species emerges from the old through the gradual accumulation of differences over long periods of time, typically in a population of one species that has become geographically separated from others of the same species. This view of evolution, the dominant one for a century and still today, is called **gradualism**.

Biologists who study the fauna (animals) and flora (plants) of isolated islands, however, have known for many years that changes in small populations can occur rapidly. Particularly where island groups like the Galapagos and Hawaii arise in the open ocean, the first animals to be blown in by a storm stand a good chance of setting off an evolutionary explosion into many new species. Similarly, when the environment, especially the climate, changes and exerts different selective pressures, as for example at the onset of an ice age, evolution can occur rapidly because of (a) the great amount of genetic variability in a species, (b) the decline of populations adapted to the old environment, and (c) the rise of populations adapted to the new. A gradual transition, one that occurs stepwise, does not necessarily have to be a slow one.

The known fossil record in Darwin's time did not offer him the great breadth of detail available to modern scientists, but it was sufficient to help shape his thinking. Though the fossil record is far better known today, many gaps remain: there are few geological deposits that extend uninterrupted for millions of years and thousands of square miles, and more information is always welcome. Recently, paleontologists Niles Eldredge and

Stephen Jay Gould have suggested that the fossil record shows long periods of very little change interrupted by brief periods of very rapid change, followed again by long periods of **stasis** (little change). They suggest that major change is related to speciation events. Each of the various species of fossil horses mentioned above persisted for several millions of years without appreciable change, and then seemed to undergo a burst of relatively rapid evolution—in perhaps tens of thousands of years, rather than millions. This manner of evolving by fits and starts, called by Eldredge and Gould **punctuated equilibrium**, has been portrayed as contrasting with the traditional gradualistic view.

Darwin frequently described evolution as "gradual" and wrote that it "generally acts with extreme slowness." Darwin was aware, however, that the rate of evolutionary change was largely affected by the rate of environmental change. He understood that in a rapidly changing environment, the tempo of evolution would be accelerated. The advocates of punctuated equilibrium and the mass media have made more of the supposed difference between punctuated equilibrium and traditional gradualism than is actually there.

Punctuated equilibrium is really just accelerated gradualism. The punctuated equilibrium proponents compress gradual changes into brief episodes. Gradual change means that the descendants are only slightly different from their ancestors. Even though there may be geologically brief bursts of evolution followed by long periods of no change, the change that occurs is still gradual, or **stepwise**. No knowledgeable biologists today, and that includes the punctuated equilibrium enthusiasts, advocate single-generation **macromutational** jumps (mutations with large effects). Amphibians, for example, did not arise from the lobe-finned fishes in a single-generation leap (**saltation**), but by seeing the muscular fins and lung of the aquatic ancestor gradually refined for terrestrial life. Macromutations do occur, but they are usually harmful (the recipients seldom produce offspring) and therefore relatively unimportant in evolution.

Gradualists maintain that the periods of stasis seen in the fossil record reflect the maintenance of the status quo by interacting species once a habitat's niches are filled; following a period of adjustment, the populations of the various plant and animal species in that habitat settle into **equilibrium**. This process is called **stabilizing selection**. Rapid change can occur when the environment changes or when extinctions vacate ecological niches, thereby promoting **radiation** into the newly available niches by the surviving species. But this is not very different from what the punctuated equilibrium advocates say, and the difference may be more semantic than biological. (What is "rapid" anyway? Ten thousand years? A hundred thousand years?)

The creationists attempt to exploit this debate by claiming that evolutionists disagree among themselves about evolution. But evolutionists debate the *tempo* and *mechanisms* of evolution, not its *existence*. The debate within evolution demonstrates the intellectual health, vigor, and usefulness of the theory, not its demise.

Much of the debate today between the punctuated equilibrium advocates and the traditional Darwinian gradualists is due to the different perspectives of the various kinds of evolutionary scientists. For example, taxonomists and population geneticists must deal with the variations present within the species that are alive today. In their studies of **subspecies** (geographical races) and **clines** (a cline is a continuous gradient of variation in a character in a population, along a line of environmental or geographical transition), they readily see how an accumulation of variations acted upon by natural selection in a different environment can ultimately give rise to a form distinct from the parental species. Paleontologists, on the other hand, must contend with the difficult problem of time. They see fossil species extending over hundreds of thousands or millions of years with very little change, but they have no way of knowing if a member of a species found early in its geological history could still have interbred with what appears to be the same species near the end of its geological reign. And although that sort of persistence can give the appearance of stasis, it is impossible to know whether or not the genes of the later populations would have accommodated the genes of the earlier populations.

Paleontologists, after all, usually have only hard parts to examine: shells, bones, teeth, etc. The soft parts of internal anatomy, and the more subtle genetic legacy of biochemistry and behavioral characters are, of course, usually not available. In making their species distinctions, the paleontologists are confined to the evidence at hand, which might encourage them to lump together over time what were, if fact, several species, thus obscuring what may have been a gradual but real change. Population geneticists and taxonomists who deal with contemporary organisms can examine more characters, and even individual genes, and often can make much finer distinctions between and within different species, than can the paleontologists.

When punctuated equilibrium advocates find an abrupt appearance of a new species and an abrupt disappearance of an old species in the fossil record, they point to this as evidence of a sudden burst of speciation. But what they see may be something quite different—migration into an area—as pointed out by Richard Dawkins in *The Blind Watchmaker*. We saw in Chapter 1 that speciation requires geographical isolation and genetic variation. If population A splits into populations A and B such that B becomes geographically disjunct from A, then population B may gradually develop into a new species through stages B_1, B_2, B_3, . . . B_{12} in the new locality. Subsequently, B_{12} may return to area A and force species A into extinction via competitive exclusion. In the fossil record of locality A, B_{12} will appear to have derived directly from A and to have rapidly succeeded A. In fact, however, B_{12} evolved gradually, in a different area. If one examines only area A, the stepwise transition from A to B_{12} will appear abrupt. If one also looks at area B, the gradual nature of the transition becomes more apparent.

It is not that punctuated equilibrium is a mistaken notion; it is just not the revolution in evolutionary theory that the press and its advocates claimed. It is certainly not anti-Darwinian. It is a great help in explaining why there should be gaps in the fossil record. Like many new ideas that have come along in the history of science, punctuated equilibrium is likely to find its place. It will be absorbed by traditional evolutionary theory to help the theory more completely explain both the diversity of life and the variety of ways and paces by which life can change.

The unity of the gradual and punctuational modes is in fact illustrated in a study by Malmgren, Berggren, and Lohmann

(*Science*, 1984, 225: 317–19). They examined a single **lineage** of planktonic foraminifera (tiny marine animals) over the last 10 million years of the lineage's history. An analysis of the changes in the lineage revealed that it was static over about 5 million years, but then underwent gradual physical change that was relatively rapid but not geologically instantaneous. The change led finally into what was clearly a new species, but not to lineage splitting. This pattern has been termed **punctuated gradualism**.

Creationists frequently point to the lack of transitional fossils as a weakness of evolutionary theory. But it is not reasonable to expect every single species that ever lived, or even very many of them, to show up in the fossil record. There will always be gaps, owing to the nature of fossilization and the chanciness of discovery. The punctuational mode of evolution, then, offers an excellent explanation for the relative scarcity of transitional forms between species. It implies that major changes can take place relatively rapidly in small populations in isolated areas, and because such transitional populations persist for a relatively short time, in a limited area, the chance of fossilization (and therefore of discovery) is very slight. Thus transitional forms between species *should* be rare in the fossil record. Such a statement is not made to protect evolutionary theory from falsification but to warn how rare such fossils are likely to be. *That even one transitional fossil is found is a sufficient demonstration of evolution and a resounding falsification of creationism*. Transitional forms are scarce at the species level because species are so much alike that it is difficult to tell, from a few bones, one species from another. Transitional forms become progressively more common as one moves up the taxonomic hierarchy, from genera to families, orders, classes, and phyla.

Transitional forms between major groups—such as *Archaeopteryx*, which was intermediate between two classes—do exist, a fact to which creationists close their eyes. We have already discussed the wealth of transitional fossils within the horse family, and there are others. *Eusthenopteron*, a likely ancestor of the amphibians (see Figure 21, in Chapter 4), was a Devonian fish with air-breathing capabilities whose fins were sufficiently robust that it could leave the water and drag itself about on land. In the Carboniferous, a group of amphibians called anthracosaurs show a mixture of amphibian and reptilian characteristics. In the Triassic, mammal-like reptiles called therapsids gave rise to the first mammals. These and other transitional forms are discussed further in Chapter 4.

In this chapter we have seen how rocks and fossils can be dated, and we have seen that these dates prove that Earth is almost inconceivably ancient. We explored the sort of information about evolution that can be derived from the fossil record. We have seen that the fossil record shows transitions between groups as predicted by evolutionary theory, and we have described the **phylogeny** (evolutionary history) of a lineage (horses). In Chapter 3 we will use the powerful theory of evolution to explain some fascinating biological realities.

3 The Explanatory Power of Evolution

A casual view of a school of minnows or a flock of blackbirds might suggest that they are all exactly alike, like molecules of water in the ocean. But they are not; many, many minor differences in size, form, or behavior will reward a closer examination. These *microevolutionary* differences accumulate in a population's gene pool over time. A few vanish, when their owners fail to reproduce, but most remain, with greater or lesser effect on the population as a whole. These differences continue to build up when populations of the same species become geographically isolated. Eventually, enough differences accumulate (through natural selection operating on the heritable variations in the gene pool) that interbreeding between the separate isolated populations is no longer possible; what had been separate populations of the same species can become two different species. This situation may go on to produce *macroevolutionary* change. A whole new line of evolution can result, depending on the new gene pools and the environmental forces acting on the new and old species; in Chapter 2 we saw this demonstrated in the fossil record of horses, and in the origin of birds from reptiles. In individual cases, the evolutionary process may take greater or lesser amounts of time, from decades or centuries to hundreds of millions of years, depending on generation time, reproductive rate, environmental change, and other factors. In what follows, we shall examine several examples of microevolutionary change, much of which can be observed today, and we will go on from there to a stepwise scenario for the origin and development of greater complexity.

Drug Resistance in Bacteria
Why Should I Take All My Pills?

When your physician gives you a prescription for an antibiotic, he or she will caution you to take the entire seven-day regimen—all the pills in the pill bottle. The physician's insistence that you not stop taking the medication after three or four days is based on good evolutionary biology. After a few days of the drug therapy you may feel better, because the antibiotic will have killed a great many of the most susceptible bacteria. But if you discontinue the drug prematurely, *some* of the bacteria will survive, and these, which are the more *resistant* bacteria, will be the ones contributing their genes to the next generation. (Bacteria, of course, produce whole new generations on a time scale of minutes and hours.) In a few more days you could be much sicker than before, and a new prescription, this time of a more potent drug, may not do as well as the first one would have. By using the entire initial prescription, then, you have a good chance of killing all of the target organisms; by using just part of it, you are, in effect, selecting for drug-resistant bacteria.

The same sort of evolutionary reasoning is applied when flu or cold sufferers, knowing only that they feel sick, beg their doctors for an antibiotic, such as tetracycline. Influenza and the common cold are caused by viruses, and whereas antibiotics kill bacteria, they are ineffective against viruses. By unnecessarily taking an antibiotic, the patient does nothing to combat the flu or cold, and instead risks selecting for a resistant strain of bacteria that is already present in the body—a strain that in the future may require even stronger drugs or greater doses to control. It should be emphasized that the antibiotic does not *cause* the bacteria to become resistant; it simply sets up the conditions that encourage the microevolutionary shift to a new strain. The genetic mutations that confer resistance on certain individuals of the bacterial colony occur independently of exposure to the drug. But it is only in the presence of the drug that the resistance shows itself, for when confronted with the drug the resistant individuals inevitably yield more descendants than the others can. That the mutations occur independently of the drug has been confirmed repeatedly in the laboratory: a number of colonies of the same original bacteria are cultured in isolation from each other, and in time some of the isolated colonies show themselves to be drug-resistant even though they were never exposed to antibiotics. More often than not, fortunately, no resistant variations are present, and the bacterial infections are totally eliminated by the drug therapy. Organisms like bacteria, which have very short reproductive cycles and huge populations, have an enormous advantage over, say, the peregrine falcon in adapting to lethal environmental changes before they are driven to extinction. Bacteria experience far more mutations because there are so many more individuals and generations. This and the short reproductive cycle allow beneficial mutations to be exploited by natural selection rapidly.

The evolution of pests that are resistant to a particular pesticide works in the same fashion. The repeated wholesale spraying of the pesticide kills the most susceptible insects and leaves the most resistant individuals to breed the next generation. The chemical companies then introduce newer, more potent poisons in an attempt to stay half a jump behind the bugs—and like bacteria, insects reproduce rapidly, in huge quantities.

Rabbits and Myxomatosis
What Are the Limits of Biological Control?

The story of rabbits in Australia is another fascinating case history in evolutionary biology. Rabbits are not native to Australia. Before their introduction, their ecological niche was filled by a great variety of small kangaroos called wallabies. In 1859, 12 wild European rabbits, *Oryctolagus cuniculus,* were imported from England. By 1886 their descendants were colonizing new areas of southeastern Australia at the rate of 66 miles a year in all directions. By 1907 the rabbits had reached both the west and east coasts of Australia, roughly the distance between California and New York. Nothing could stop the plague of rabbits. Thousands of miles of "rabbit-proof fences" failed to stem the tide. Certainly the wallabies had offered no competitive resistance, and the few native predators made scarcely a dent in the rabbit populations. Hunting, trapping, and poisoning were to no avail. The rabbits were eating much of the sparse vegetation that supported Australia's huge sheep and cattle industry, and the graziers were suffering enormous financial losses.

The only solution was biological control. After much testing, government biologists introduced a mosquito-borne virus called myxomatosis. This virus caused a nonlethal disease in its natural host, a South American rabbit, but the disease was deadly for the European rabbit and completely harmless to all other Australian wildlife, domestic animals, and humans. To all indications, the solution had been found.

The disease did indeed take hold in 1950, and by 1952 it had produced a nationwide epidemic in the rabbit population. The mortality rate reached 99.9 percent, *but a good evolutionary biologist could predict what would happen next*. A parasite that invariably kills its hosts before ensuring its own survival would be selected against (all of its individuals would die). And that is what inevitably happened with the myxomatosis disease. The viruses had been randomly mutating, and the mutations that produced less virulence were selected (because the more virulent strains died with their hosts). The rabbits too, were mutating, and they were being selected for greater resistance to the disease. The result was a milder disease and stronger rabbits—therefore more rabbits. Today the mortality rate is down to about 40 percent. There are still annual outbreaks of myxomatosis in Australia, but the disease is less effective in controlling the rabbits. This is evolution in action, instigated and observed by humans, and occurring through natural evolutionary forces: *it is not explainable by any other concept*. Meanwhile, the government biologists are trying to develop more virulent strains of myxomatosis. And so it goes.

The Peppered Moth and Industrial Melanism
Can Even Air Pollution Drive Evolution?

One of the most celebrated cases of evolution via natural selection is the shifting fortunes of the peppered moth, *Biston betularia*, observed steadily by scientists for 140 years. This English moth exists in two distinct color phases, light and dark (Figure 13), and individuals of the one phase routinely and successfully mate with individuals of the other phase (thus they are not separate species). Only one pair of genes is involved in the color differences, and dark is dominant to light. (A dominant gene manifests its full effect despite the presence of a contrasted (recessive) gene whose expression for the character is blocked). The moths typically rest on lichen-covered tree trunks and branches, and their main predators are birds. Museum collections made in 1848 (prior to the Industrial Revolution) indicate that the frequency of the dark form was at that time less than 1 percent of the total of peppered moths in Manchester. While resting on the lichens, which are light, the dark variant was clearly visible and was consequently easily spotted by the birds that feed on the moths. The light variety of the moth, far better camouflaged against the lichens, was much less noticeable to the birds, much less often eaten, and far more often passed its genes along.

Fifty years later, the Industrial Revolution, with its sooty air pollution, had blackened and killed the lichens growing on the trees. The dark-colored moths now made up about 95 percent of the population. Films of feeding birds show them selectively eating the conspicuous lighter-colored moths. But in the 1950's, stringent anti-pollution laws were passed in Britain, and since then the air quality has greatly improved and soot has been reduced. *As predicted by evolutionary theory*, the white form is increasing its numbers once again. Natural selection (differential reproduction), brought about by the birds' inevitable concentration on the more visible form, resulted in a change in gene frequency (evolution). Speciation is not involved, since both light and dark forms are the same species. The existence of a dark-colored variant of a species is called **melanism** (there are

Figure 13. The peppered moth, *Biston betularia.* At left, light and dark (melanistic) forms rest on a lichen-covered tree in an unpolluted countryside; at right, both forms rest on a soot-covered tree near an industrial area of Birmingham, England. A light-colored moth on a light-colored tree (or a dark on a dark) is said to be cryptically colored, or disguised. Birds exert selective pressure on the moths by feeding chiefly on the noncryptically colored form (the one they more easily see) in each environment (see the text). (Photograph by H. B. D. Kettlewell; reproduced by permission from David Kettlewell.)

black-squirrel variants of gray squirrels, and many other such examples); the shift to dark variants of the peppered moth at the height of the Industrial Revolution has been called industrial melanism.

Sickle Cell Anemia and Malaria
How Can Bad Be Good?

A case that most elegantly exemplifies the explanatory power of the principle of natural selection in the theory of evolution is the relationship between sickle cell anemia and malaria. Sickle cell anemia, an often devastating hereditary disease of many African and Mediterranean people, results from the body's manufacture of defective **hemoglobin**. Hemoglobin is the complex molecule that carries oxygen in the red blood corpuscles. It consists of four intertwined polypeptide chains—two identical alpha chains and two identical beta chains. The two alpha chains contain 141 amino acids each and the two beta chains are composed of 146 amino acids each, for a total of 574 amino acids in the molecule. The only structural difference between the normal hemoglobin molecule, called hemoglobin A, and the sickle cell molecule, called hemoglobin S, is one change in the sequence of each beta chain: the amino acid valine is substituted for glutamine. That single substitution in the two beta chains, out of 574 component amino acids, yields profound effects: sickle cell hemoglobin molecules distort the red blood corpuscles. Normally disk-shaped, the corpuscles become rigid and sickle-shaped (see Figure 14), and these abnormal cells tend to clog the smaller blood vessels. When the body attempts to destroy these defective cells, anemia results. Sickled cells may have a life span of only 30 days vs. the 120 days of a normal red blood corpuscle, and infants with sickle cell anemia are highly sensitive to infection.

The amino acid substitution in the beta chain that produces the sickle cells is under genetic control. Humans, like other mammals, receive one set of genes from each parent. (It may help to review the Appendix at this point.) Normal hemoglobin is dictated by the gene Hb^A, and most people have two of these genes

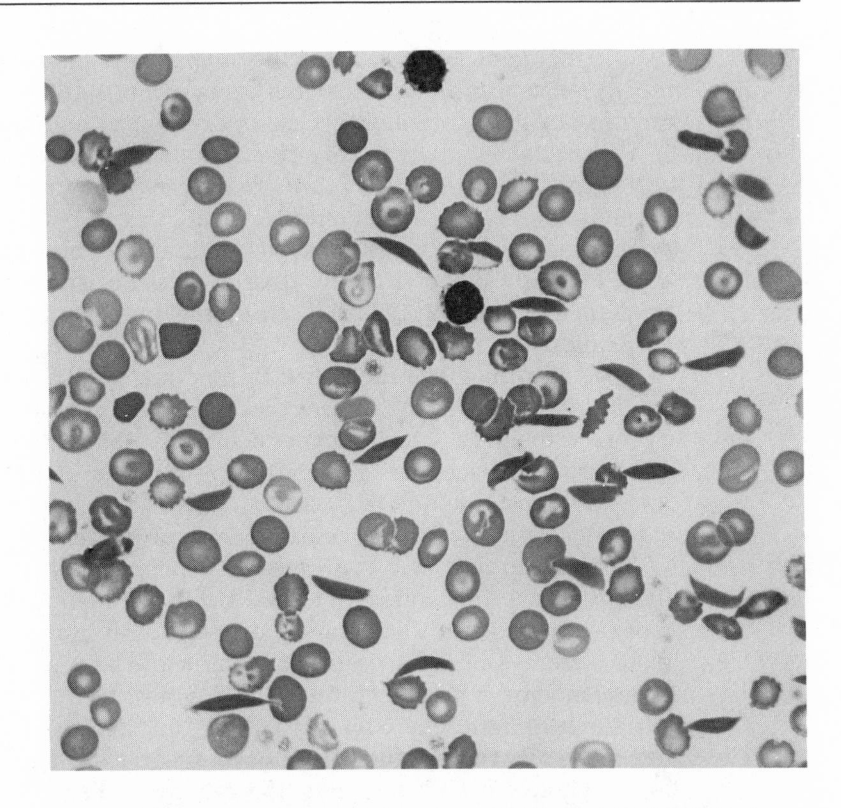

Figure 14. Human blood smear, showing normal and sickled red blood corpuscles. Normally, oxygenated red blood corpuscles are disk-shaped, but under low oxygen conditions, the red blood corpuscles from an individual with sickle cell trait (only one sickle cell gene) distort and cannot readily pass through the body's capillaries. The resulting blood clots are painful and reduce blood flow to various organs, resulting in illness and often a reduced lifespan. (Photo courtesy of Carolina Biological Supply Co.)

(Hb^AHb^A). Persons who have two mutant alleles (an allele is an alternate form of a gene), Hb^SHb^S (one Hb^S inherited from each parent), have sickle cell disease and are severely affected; most die in childhood. People who are heterozygous, Hb^AHb^S (they inherit one normal gene and one mutant gene), have sickle cell *trait*. Their bodies produce both normal and sickle cell hemoglobin; neither gene is dominant to the other. These people can live mostly normal lives, and generally they show signs of anemia only in stressful conditions, such as exertion at high altitudes.

The sickle cell allele is very common in the peoples of west and central Africa (Figure 15). Its frequency is 15 to 20 percent in many areas, and, in some tribes, as high as 30 percent of the population have sickle cell trait. The evolutionary biologist must ask, "How is it that a virtually lethal allele can exist at such high frequencies? Shouldn't natural selection eliminate a lethal allele very quickly?"

If we plot the geographic distribution of the sickle cell gene, we find that the area in Africa where it occurs corresponds quite closely to the range of malignant falciparum malaria (Figure 15). This virulent disease is transmitted by the bite of the female *Anopheles* mosquito and is caused by a protozoan parasite, *Plasmodium falciparum*, that attacks red blood corpuscles.

Medical studies have confirmed that people with sickle cell trait (that is, with both normal and sickle cell hemoglobin) are more resistant to malaria than those with only normal hemoglobin. Here we have a beautiful example of Darwinian theory. Where malaria is prevalent, the heterozygotes, Hb^AHb^S, are actually superior in fitness to the normal Hb^AHb^A. Because the defective allele Hb^S protects its carrier from malaria, and the normal homozygote Hb^AHb^A does not, the defective alleles are maintained at a high level relative to the normal genes. Malaria, then, is a greater risk to life in this region than is sickle cell trait. The type of stabilizing selection demonstrated here is known as a **balanced polymorphism**.

Blacks were, of course, kidnapped from west-central Africa and shipped into slavery in the American markets until well

Figure 15 (facing page). Old World distribution of the sickle cell gene, Hb^S (upper map; in percent of population) and of malaria (lower map). If two sickle cell genes are present in an individual (the homozygous condition), sickle cell anemia results, and it is usually fatal. But if only one sickle cell gene is present (the heterozygous condition), the individual is said to have sickle cell trait, which, under normal conditions, is neither debilitating nor lethal. Falciparum malaria (lower map) is caused by the protozoan parasite

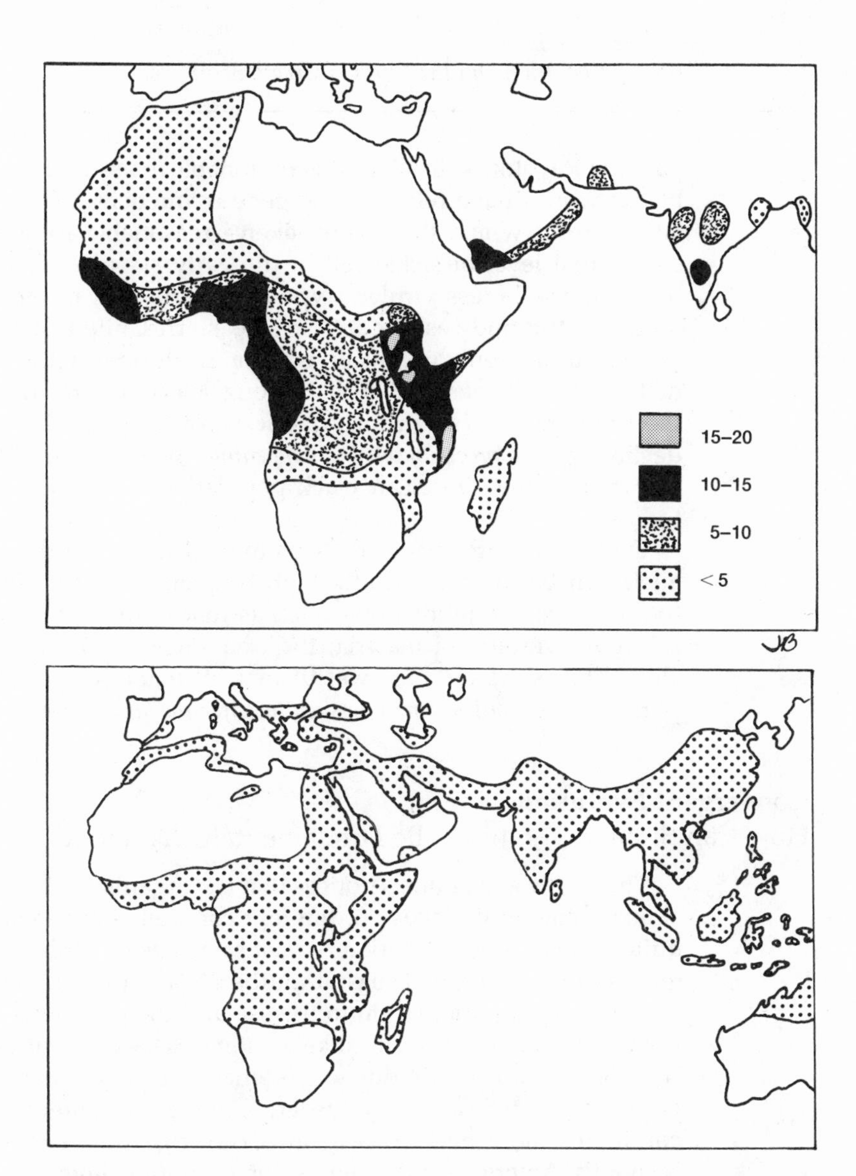

Plasmodium falciparum, which is transmitted by the bite of the female *Anopheles* mosquito. The frequency of the sickle cell gene tends to be higher where falciparum malaria is endemic, because people with one sickle cell gene gene have a greater resistance to malaria than those with normal hemoglobin (Hb^AHb^A). This is an excellent example of how a deleterious gene, which in other circumstances would be selected against over time, can be kept at a high frequency indefinitely in a population.

past the Revolutionary War. It is reasonable to assume that these blacks would have reflected the gene frequencies of the populations from which they were taken—reasonable to assume a substantial level of sickle cell in the early slave population. It happens that a less virulent type of malaria was formerly epidemic in the midwestern and southeastern United States, and was eradicated only relatively recently. Evolutionary theory predicts that in the absence of that strong selective force (malaria), the frequency of Hb^S in the American black population should decline. *This is precisely what is happening*; the frequency of sickle cell trait in the American black population is about 8 percent today.

Evolutionary theory enables us to explain how a deleterious allele can be maintained at a high frequency, and it allows us to make testable predictions, such as that in the absence of the selective pressure of malaria, the frequency of Hb^S should decline. This is the sort of powerful analysis that a scientific theory permits. *Creationism offers no such tools for the analysis of reality.*

Convergent Evolution
How Can Similar Organisms Be Fundamentally Different?

The alert traveler often notices that the organisms that occupy similar ecological niches on different continents are themselves quite similar. The vegetation of vast stretches of Chile instantly recalls the vegetation of coastal southern California: in climate, in geology, and in orientation to the sea and the prevailing winds, the two coastlines are quite alike. In some cases these similarities between organisms are due to descent from common ancestors. For example, the cat family has large, powerful hunters on most continents (lions and leopards in Africa, tigers in Asia, jaguars in South America, and cougars, or mountain lions, in North America). No one doubts that the big cats are all closely related.

But in other cases, quite unrelated organisms, when subject to the selective pressures of the same ecological niche, or of similar niches on different continents, have often evolved a whole complex of similar adaptations, including body form and behavior.

The process driving these cases is called **convergent evolution**, and the accumulated similarities can be quite remarkable.

The wolf and the thylacine, both of them large hunting carnivores that run down their prey on open ground, serve to illustrate the process. There are wolves across the Americas, Eurasia, and Africa, but in Australia, which has been isolated from the other continents for 50 million years, there are no wolves, not even fossil wolves. In fact, the vast majority of Australian mammals—koalas, wombats, kangaroos, thylacines, and scores of others—are marsupials: the young are born soon after conception and must remain in a pouch on the mother's belly for weeks or months. We now know that Australia, with its marsupial cargo, drifted away from the other continents before the more recently evolved placental mammals (except bats) could reach it. (In the placentals—virtually all of the world's mammals outside of Australia—the growing embryo is fused to the wall of the uterus and enabled to grow to considerable size before birth.) Although marsupials originated in South America in the Cretaceous, in the era of dinosaurs, today South America has fewer species (78) of marsupials than do Australia and New Guinea (172); and North America has only one, the opossum. Marsupials reached Australia when South America and Australia were connected via Antarctica. Competition from placental mammals eventually reduced the number of marsupial species in South America, whereas Australia's isolation as an island allowed the marsupials to radiate into many new species.

One of these remarkably diverse animals was *Thylacinus cynocephalus*, a large hunting carnivore generally called the Tasmanian "wolf" or "tiger," or simply "thylacine." It has a wolflike body, feet, ears, and hunting behavior, and the powerful jaws of the wolf, but its similarity to the wolf ends there. Like the females of its many marsupial relatives, the thylacine female rears her young in her pouch after their "premature" birth, and the brain of *Thylacinus* is small, only 40 percent as large as a wolf's brain. Sadly enough, these animals are probably extinct, having failed to compete with the more intelligent dingo, Australia's wild dog. The dingo, of Asian wolf ancestry, was introduced by the Aborigines about 3,500 years ago, and its effect on the

Australian fauna was no doubt immediate and devastating. The European settlers, after their arrival in 1788, slaughtered thousands of dingoes and practically all of the remaining Tasmanian wolves, because of their predation on sheep and other domestic animals. The last known Tasmanian wolf died in the Hobart, Tasmania, zoo in 1936.

Australia also has a marsupial mole, *Notoryctes typhlops*, which resembles the true moles of the Northern Hemisphere in both behavior and morphology. Natural selection has enlarged the forelimbs and markedly reduced the eyes of the marsupial mole, just as it has the forelegs and eyes of its distant placental counterparts; both of these developments are adaptations for burrowing through the soil, in darkness. The marsupial glider *Ptaurus breviceps*, often called the flying phalanger, has a broad fold of skin along each side of the body, from wrist to ankle, and a bushy tail, just like American flying squirrels, which are placental mammals (rodents). Both leap great distances from tree to tree using the skin fold as a parachute. Among the other Australian marsupials are counterparts of placental cats, lemurs, mice, and anteaters (see Figure 16).

In the sea, sharks (cartilaginous fishes), swordfishes (bony fishes), dolphins (mammals), and the long-extinct ichthyosaurs (reptiles) all have similar streamlined, efficient torpedo shapes, tall dorsal fins, and broad tails. All are (or were) fast and agile marine predators, but their ancestors were just remotely related.

In the southern African deserts are succulent plants of the spurge family, Euphorbiaceae, that look for all the world like the cacti of the Western Hemisphere, but have wholly different ancestors. In both cases, there are tall spire-like species, barrel-shaped species, shrubby species, ground-hugging species, and

Figure 16 (facing page). Convergent evolution in selected Australian marsupials (righthand column) and their counterparts among the placental mammals of the rest of the world (lefthand column). The remarkable similarities reflect evolutionary convergence from fundamentally different stocks, and the power of natural selection operating in comparable ecological niches. (Reproduced from P. H. Raven and G. B. Johnson, 2nd ed., 1989, *Biology*, by permission from C. V. Mosby Co., St. Louis.)

MOLE
MARSUPIAL MOLE
ANTEATER
NUMBAT (ANTEATER)
MOUSE
MARSUPIAL MOUSE
LEMUR
SPOTTED CUSCUS
FLYING SQUIRREL
FLYING PHALANGER
BOBCAT
TASMANIAN "TIGER CAT"
WOLF
TASMANIAN WOLF

still others, each adapted to a particular desert ecological niche, but their floral parts and other basic features show that the American cacti (family Cactaceae) and the South African euphorbs are unrelated.

Such close similarities in very unrelated groups are easily explained as a result of convergent evolution. The environment simply works with what it has, and in due course selects the most efficient design for the animal's or plant's lifestyle in the particular set of environmental circumstances.

Stepwise Adaptation
Can an Accumulation of Minor Adaptations Lead to Major Changes?

For the nonbiologist, one of the most difficult aspects of evolution to grasp is the creative power of natural selection. The old argument from design says that if you find a watch, there must have been a watchmaker—a creator. And it is tempting to transfer this analogy to organs or organisms. But the "design" or order in physical, chemical, and biological systems is not the sudden brainstorm of a creator, but an expression of the operation of impersonal natural laws, of water seeking its level. An appeal to a supernatural explanation is unscientific and unnecessary—and certain to stifle intellectual curiosity and leave important questions unasked and unanswered.

But if evolution is such a mindless, unplanned process, how could such a complex organ as an eye evolve? How could so many thousands of different organisms show such exquisite adaptations for peculiar life styles?

The answer to both questions is that the sieve-like action of natural selection is creative. Natural selection doesn't have to start over from square one each new generation. It retains favorable variations that occur through mutation, and the unfavorable variations fall through the sieve to oblivion. Mutations are not induced by the needs of the organism; they occur at random. They are not designed; they just happen. And for every favorable mutation, there are probably many unfavorable mutations. But the favorable variations tend to accumulate and to lead to increasing complexity in the pre-existing structures. *Evolution is a tinkering process, not a designing process.*

The key to understanding the process of evolution is to remember that natural selection works in **stepwise** fashion. An eye didn't just appear full-blown. It evolved stepwise, one small, modest step after another, over a great deal of time. In fact, eyes have evolved several times independently—in arthropods (spiders, crabs, insects, etc.), in mollusks (snails, scallops, octopuses, etc.), and in the vertebrates. The eyes of these three groups are built differently and function differently, but they are eyes nonetheless. Very likely, eye evolution began with a spot of light-sensitive nervous tissue. This sort of proto-eye is found in some primitive animals alive today. Arranging several of the sensitive cells along the inner surface of a cup-like depression would enable the developing eye to detect the direction of the light source. If the edges of the cup nearly close up, a pin-hole camera-eye has evolved. A retina with rods that detect motion and cones that detect color is a further improvement. And if evolution then augments that with an iris diaphragm, which controls the admission of light, then even more survival benefits accrue. Develop a lens for accommodation and focusing, and the evolving eye becomes still more useful. Such stepwise changes, the results of mutations, are built up over time. No foresight is involved. Those individuals possessing any slight stage of the change may be slightly better adapted to the environment than those lacking that refinement. If so, they will tend to leave more offspring, which in turn retain the potential for further modification.

Creationists frequently make the specious argument that an eye (or ear, wing, lung, etc.) could not have evolved because the intermediate stages would be imperfect and therefore not functional. They miss the point that a structure need not be in final form to confer an advantage. Some vision is better than none. For example, if you have ever had your eyes dilated for a vision exam, you know that the iris diaphragm becomes nonresponsive to light intensity and the pupil remains wide open. Your eye is imperfect until the effects of the drops wear off. But you can still see well enough to get around, and you are certainly better off than someone with no vision. Some of us are very nearsighted.

We myopic types have an elongated eyeball that causes the light rays to converge in front of the retina instead of on it. This results in blurred vision of distant objects. But if I were a wild animal, my nearsighted vision would be a big improvement over no vision. I could see the forms of food, shelter, mates, and predators, and detect if they were moving. *Eyes did not arise suddenly from nothing*. They evolved gradually over hundreds of millions of years by incremental improvements over previous models.

The creativity of natural selection involves the retention and subsequent stepwise refinement of variations that yield improved **fitness**. Fitness in the Darwinian sense means reproductive fitness—leaving at least enough offspring to spread or sustain the species in nature indefinitely; it supposes a favorable relationship between an organism and its environment, a relationship that results in the optimization of the species' reproductive potential. But selection for fitness is certainly not a random process. Creationists misrepresent evolution when they say biologists claim an organ or organism arose by chance. No responsible biologist says such a thing. *Natural selection is the antithesis of chance*. A mutation does arise by chance, in the sense that it is an unplanned event unrelated to the needs of the organism, but the effect it has on the coming generations depends on its survival value to the individual in a given environment.

An organism's environment sets certain "problems" that the organism must "solve" in order to survive. Evolution via natural selection is the mechanism by which the organism solves the environmental problem. Just as a key is exactly fitted to its lock by a slow, careful, stepwise process of cutting and filing, natural selection fits an individual to its environment, fine-tuning the members of the population to the role they play in nature. A single stepwise result may be all but imperceptible, or it may be the beginning or end of a truly remarkable adaptation.

But natural selection doesn't produce the best *imaginable* adaptation; rather, it works with what it is given in a historical sense, with organisms imperfectly suited to a changing environment, and it works with time to produce the best *available* adaptation. *Natural selection is a remodeling process*. An excellent example of this is the human spine. It is a compromise between the needs of our four-footed ancestry and those of our own bipedal nature. An engineer could certainly design from scratch a more efficient and more pain-free backbone, but given our primate starting point, natural selection did the best it could to achieve a workable compromise. *Natural selection never starts from scratch.*

By contrast, the modern great white shark is only modestly more refined in its design than its ancestors of 100 million years ago; the tinkering process, which even then had already been at work on the variations of sharks for many millions of years, had produced a matchless predator, and because the selective forces of the open ocean have changed so little in all this time, there has been little need for further modification.

All change is not necessarily adaptive, and environments themselves change. Some changes may be simple by-products of other changes and may become adaptive much, much later in their own right, when the organism confronts a different environment. The wings of insects, for example, are thought to have been a by-product of a structure that had originally evolved for courtship display or for aquatic respiration. The feathers of birds evolved as insulation, but they paved the way for flight. The next chapter explains how natural selection has made use of such **preadaptations** in a changing environment.

4 The Evolution of Life and the Rise of Humans

For the creationists, the task of explaining the origin of life is dazzlingly easy: it was simply created. For scientists, who look to hard evidence, observable facts, and critical thinking for explanations, the task is not so easy—complexities, subtleties, and the immensities of time intrude. In seeking to recount the origins of life and the rise of humans I find myself in the awkward position of attempting to compress 15 billion years of evolution into a few thousand words. Necessarily, then, my account will hit only selected high points. This ultra-truncated view of evolution from the Big Bang to humans is designed not so much to convince anyone of the occurrence of these processes or events, but rather to offer a succinct, clear overview of the panorama of life, to suggest how scientists seek answers to questions, and to whet the reader's appetite for more information. I will include key words and fossil names so that the reader can pursue topics of interest in the library. Further readings are suggested at the back of the book.

The Big Bang
How and When Did It All Begin?

It is easy to decide where to begin. **Cosmologists** (astrophysicists who study the Universe as a whole) postulate that the Universe originated in a gigantic explosion called the **Big Bang**.

The Universe began as an infinitely hot point of infinite density, which cooled and diffused as it exploded outward. Science can make no statement about the nature of the Universe prior to that explosion, because the physics of the Big Bang is not yet fully understood. We know that space, time, matter, and energy existed after the Big Bang. Einstein's 1916 theory of general relativity, showing that matter and energy are interchangeable, and that space and time are a continuum, forms part of the framework of the Big Bang theory.

Why do cosmologists think there was a Big Bang? Practically everything we know about the Universe is based on an analysis of electromagnetic radiation such as x-rays, radio waves, light, etc., all of which travel at the speed of light. These data provided the key to the Big Bang theory by showing that the Universe is expanding. In the 1920's the American astronomer Edwin Hubble, while spectroscopically analyzing light from distant galaxies, observed that the light from objects apparently moving away from the observer shifted toward the red end of the spectrum. Light from objects seemingly moving toward the observer underwent a compression in wave length toward the blue end. This shift is analogous to the sudden drop in pitch of a horn from a speeding car as it approaches and then passes. Further observations showed that most galaxies appear to be receding at great speed, and that only a very few nearby galaxies show blue shifts. Everything is receding from everything else like spots painted on the surface of an inflating balloon; but there is no center of expansion. Hubble eventually demonstrated that the extent of the red shift shown by a galaxy is directly proportional to the galaxy's distance from us. Calculations based on the rate at which the stars seem to be receding from each other, and allowing for acceleration due to gravity, indicate that the Big Bang occurred sometime between 10 and 20 billion years ago. Recent studies of star spectra support an age of around 11–12 billion years. Most cosmologists favor a figure of around 12 to 15 billion.

In 1964 two astronomers, Arno Penzias and Robert Wilson, working at Bell Laboratories in New Jersey, detected cosmic microwave background **radiation** by using signals reflected from the Echo satellites. Cosmologists agree that this radiation is the primeval light and heat from the Big Bang itself. Their research was honored with a Nobel Prize in 1978.

The big question in cosmology today is whether the Universe will continue to expand forever, or whether it will begin to contract and collapse into the infinite density from which it originated. The answer depends on how much matter exists in the Universe. If the total amount of matter left over from the Big Bang exceeds a certain critical mass, gravitational forces will eventually bring about a collapse. If the matter in the Universe equals or is less than the critical mass, the Universe will expand forever. The calculations of the total amount of matter in the Universe are maddeningly near the border of the critical amount. The answer continues to elude us.

The Early Earth
What Was It Like Before Life?

After the Big Bang, matter aggregated into glowing masses (stars) that clustered together via mutual gravitational attraction into assemblages of stars (**galaxies**) scattered across space. Our own island of stars is the Milky Way Galaxy. It is made up of some 100 trillion stars (that's 100 followed by nine zeros) scattered across a disk roughly 100,000 light-years in diameter (it takes light 100,000 years to travel across the Milky Way). The disk of stars is slightly over 2,000 light-years thick. Nearly all the stars we see with the unaided eye are in the Milky Way.

Eventually, our rather ordinary solar system formed in the Milky Way Galaxy about 4.5 billion years ago (BYA). It condensed out of a cloud of dust created by the Big Bang. One can observe stars forming today in the same manner in the Orion Nebula. The planets, including Earth, formed out of the aggregated matter from the dust cloud around the Sun. For eons and eons, water would have existed on Earth only as superheated steam that rose to the upper atmosphere, condensed, and fell as rain, which then became hot vapors again. By 3.8 BYA at most, the Earth had cooled sufficiently to allow rock formation, since that is the earliest date reflected by the radiometric dating of Earth rocks. Moon rocks brought back by Apollo 11 and Apollo 12 proved to be about 4.5 billion years old. Eventually the Earth's surface would have cooled sufficiently to allow standing bodies of water to accumulate. These pools would offer sites for chemical evolution.

The early Earth rocks suggest an oxygen-free atmosphere (today almost one-fifth of the atmosphere is free oxygen). The presence of great amounts of ferrous iron in the early banded-iron formations indicates that free oxygen was unavailable in the atmosphere. (If oxygen had been present, the iron would be in the ferric state—that is, it would have oxidized, or rusted.)

Gases escaping from the hot interior of the Earth, through volcanic vents, may have produced an atmosphere composed of some or all of the following gases: carbon monoxide, carbon dioxide, nitrogen, hydrogen sulfide, ammonia, methane, hydrogen, and water vapor. Geochemists are currently debating the relative amounts of these gases that may have been present in the early atmosphere, but free oxygen was not one of them. Carbonate minerals in early rocks indicate that these rocks formed in an atmosphere of carbon dioxide rather than methane.

Origin-of-Life Experiments
Can Science Create Life?

One of the first scientists to suggest that life originated from inanimate chemicals was the Russian biochemist A. I. Oparin, whose pioneering work with **coacervate droplets** (solutions of oppositely charged colloids such as gelatin and gum arabic) was done in the 1920's. Coacervates absorb water to form a surface "membrane" that creates tiny packages in which chemical reactions may take place. Oparin's work, combined with that of the British biologist J. B. S. Haldane, greatly influenced research on chemical evolution and the origin of life during the 1930's and 1940's.

In 1953 a very simple, yet elegant, experiment was suggested by Harold Urey and conducted by Stanley Miller at the University of Chicago. Miller put a gaseous mixture of methane, ammonia, hydrogen, and water vapor in a flask and subjected it to an electrical discharge for a week. The results were startling. When the brown scum that formed in the flask was analyzed, many organic compounds were found that occur in living organisms. *He had artificially synthesized four amino acids, urea, and several fatty acids*. This event stimulated a great deal of research by other investigators, who found that they could create many organic compounds using a variety of gas mixtures and such energy sources as heat (to simulate terrestrial temperatures), ultraviolet light (to simulate solar radiation, because in the Earth's infancy there was no oxygen atmosphere to absorb it), and electrical discharge (to simulate lightning). It is now recognized that Miller's combination of ingredients is not necessary for the origin of amino acids and other **prebiotic macromolecules**. Similar results can be obtained from highly reactive products of the Earth, such as cyanides or formaldehydes, or from an atmosphere such as carbon monoxide, nitrogen, and water vapor. Hydrogen cyanide can easily be formed from a mixture of ammonia and methane in the presence of electrical discharge. Although cyanide is a poison to oxygen-requiring life, it may have been central in prebiotic chemistry because of its ability to form larger complexes.

Simulation experiments have demonstrated how easily the basic compounds could have formed. These, in turn, could give rise to many other possibilities, such as components of nucleic acids and adenosine triphosphate (ATP), which is the energy system of all organisms. There is no doubt that all of the building blocks of life can be accounted for with the various simulation experiments. The stage was now set for the origin of life.

Research by Sidney Fox, at the University of Miami, and others provides a scenario for the origin of living cells. By heating (to less than 150°F) combinations of amino acids, including aspartic or glutamic acid (these amino acids are found in simulation experiments, in the field, in lunar soil, and even in meteorites), Fox was able to **polymerize** (combine) the amino acids. Such a reaction could occur on a sun-warmed surface. An important discovery was that the amino acids do not polymerize randomly. They are self-ordering and form highly specific sequences, which greatly enhances the likelihood of such a reaction. When these polymers later come in contact with water they form remarkable structures called **proteinoid microspheres**. Rain could induce this reaction, and would very likely wash the resulting microspheres into the sea, where further development could occur. The microspheres possess many of the characteristics of the original polymers, plus emergent properties unique to the microspheres. Proteinoid microspheres have a double-layered, electrically active surface resembling a membrane. If the microspheres are left undisturbed in their solution, they can "grow" by absorbing more proteinoid material. They move, form buds like those of yeast or bacteria, and may even divide. Microspheres also show osmotic and selective diffusion properties, allowing small molecules to pass through their boundaries while retaining large molecules. These properties are shared with living cells and could not be predicted from the original characteristics of the component molecules.

Under a microscope, the microspheres look like primitive cells. In fact, artificially fossilized microspheres are indistinguishable from the earliest known microfossils that date back to about 3.5 BYA (Figure 17). I asked Dr. Fox if he would be willing to say that his proteinoid microspheres were alive. He responded that, "They are protoalive." That is not an evasive answer but a reflection of the difficulty of defining life. For centuries, science knew nothing intermediate between non-living and living things, *but today the distinction is not at all clear*. Since life evolved from non-living matter, at some point we must arbitrarily draw a line and say that everything past that point is alive. Viruses, for example, appear to be alive when they infect a host, but seem to be non-living when outside a host. Viruses cannot reproduce or metabolize on their own. They require the host's help. Things that metabolize, reproduce, respond to stimuli, and adapt are usually described as being alive. Viruses are on the borderline between life and non-life. *A single cell seems to be the smallest unit that can be said to be alive*. Proteinoid microspheres may be called **protocells**.

In modern cells, **DNA** provides the coding mechanism for the production of **enzymes** (proteins that enhance the rate of cellular reactions; see Appendix A) that run the cells. In the

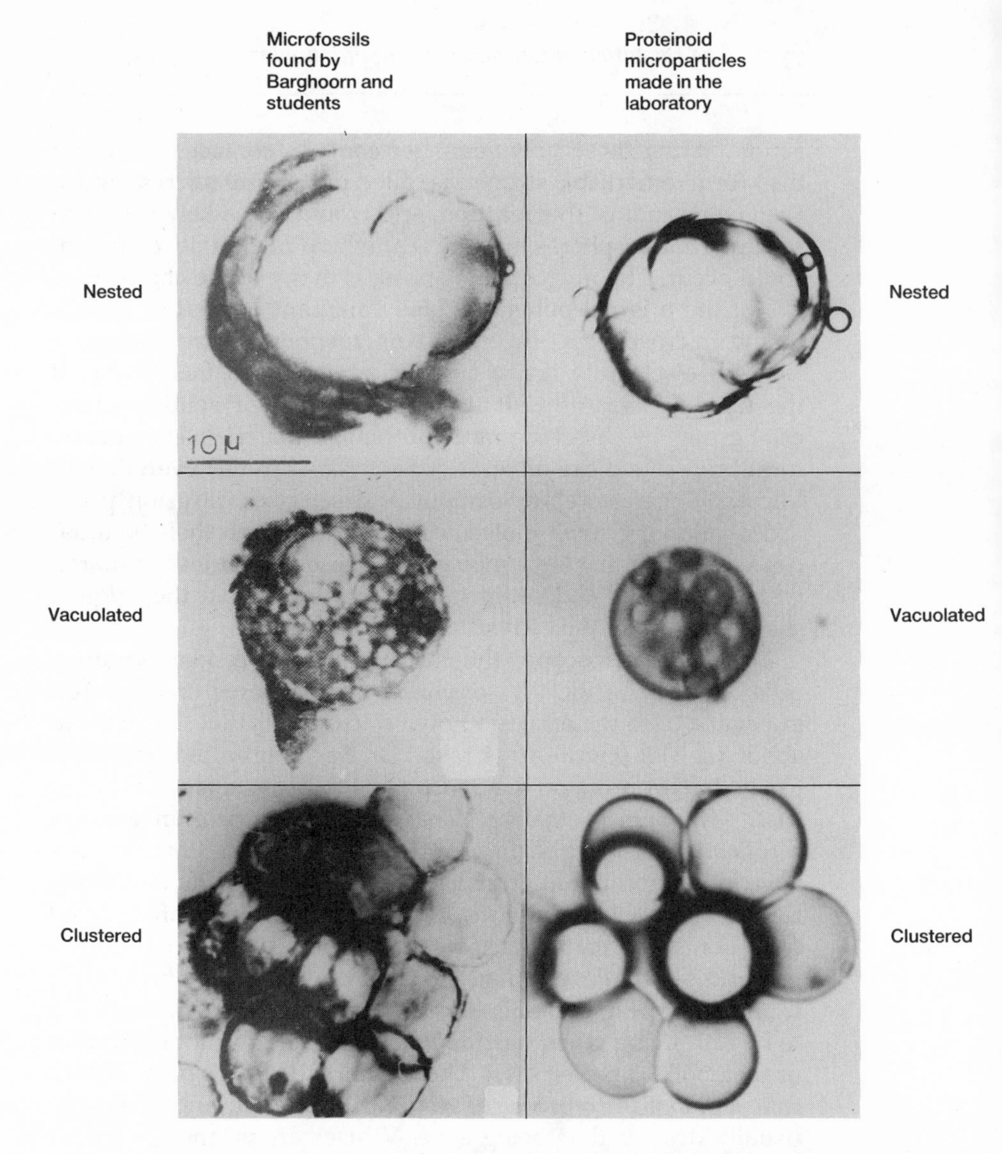

Figure 17. Microfossils and microspheres. Microfossils found by Barghoorn and his students in Precambrian rocks (left column) are strikingly similar to laboratory-made proteinoid microspheres (right column). Nested (top), vacuolated (middle), and clustered (bottom) forms of both are shown. (Photograph courtesy of Dr. Sidney W. Fox, from S. W. Fox and K. Dose, 1977, *Molecular Evolution and the Origin of Life*, rev. ed., by permission from Marcel Dekker Inc., New York.)

foregoing, note that there was no mention of nucleic acids in the formation of the protocells. The reactions that yield protocells do not require RNA or DNA. If life originated through the proteinoid microsphere stage, then protein preceded RNA and DNA. This solves the chicken-or-egg problem of which came first. The main difference between protocells and cells is that cells possess a **genetic code** based on nucleic acid, whereas protocells do not; the origin of a genetic mechanism was a later development. Protocells (using lysine-rich proteinoids as a catalyst) have been shown to be capable of synthesizing small polynucleotides, which are some of the ingredients for a DNA coding mechanism. Protocells that developed more efficient biochemical pathways, such as ATP, light-active enzyme systems, coding mechanisms, etc., would be selectively favored over less sophisticated protocells. Protocells thus provide for three of the most fundamental problems in the origin of life: the ordered self-assembly of proteinoids; the emergence of membranous structures from nonmembranous components; and the appearance of enzymes in the absence of enzymes to make them. We are very complex organisms; but everything in us, including what constitutes a brain, is built of pretty ordinary stuff.

Development of an Oxygen-Rich Atmosphere

Has Air Always Been Air?

Eventually, organisms like cyanobacteria (blue-green algae) evolved. These primitive microbes first appeared in the fossil record over 3 BYA (Figure 18). They were capable of using water as a hydrogen source in manufacturing nourishment from sunlight and inorganic chemicals. This process, called **photosynthesis**, liberated oxygen into the atmosphere, and by 2 BYA the accumulation of oxygen had become substantial. This shift in the composition of the atmosphere was a momentous development, and the evidence for it is in rock formations that could only have been laid down in the presence of atmospheric oxygen. Oxygen eventually came to shield the Earth from intense ultraviolet radiation, thus allowing organisms to colonize shallow water. Oxygen may also have been lethal to some of the primitive microbes that had evolved in its absence. By contrast, it was the organisms that evolved various oxidation reactions for the synthesis of ATP that were favored by selection.

Bacteria and cyanobacteria lack a well-defined **nucleus** (cell

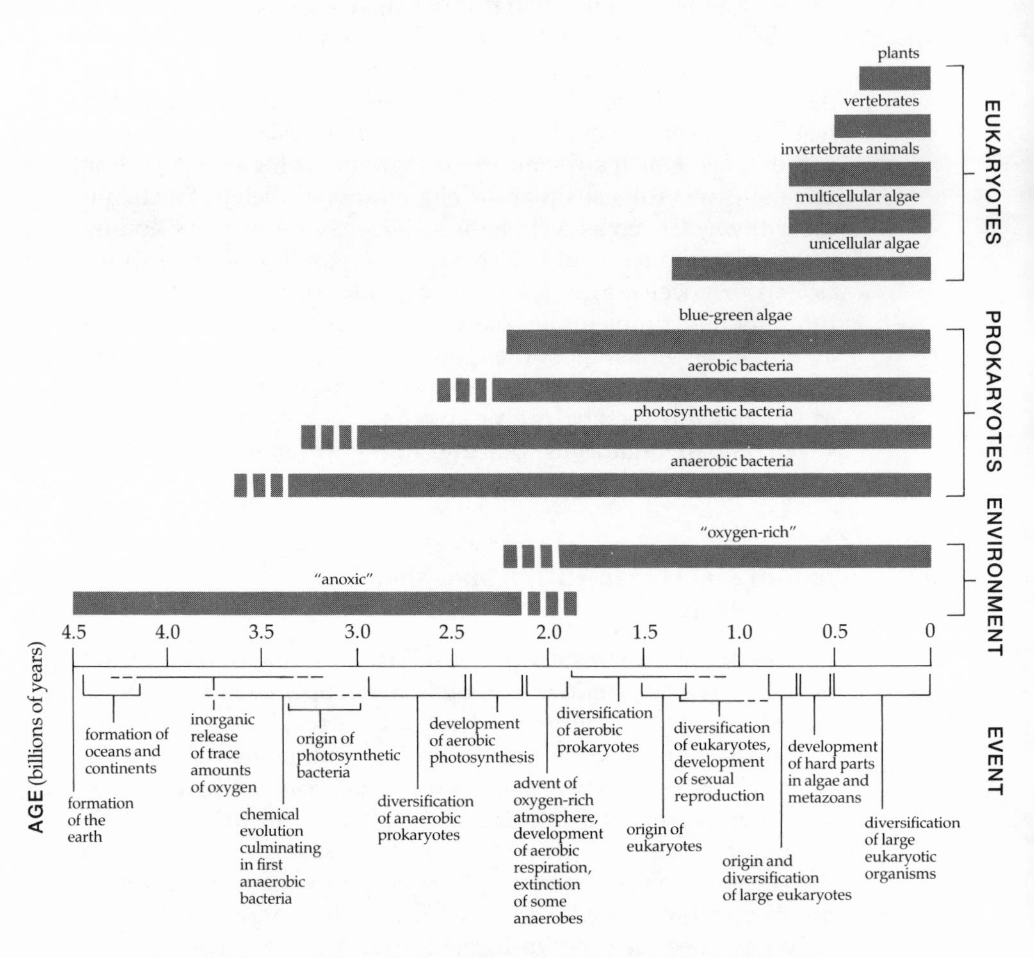

Figure 18. Chronology of the major events leading to the evolution of ancient and modern organisms. (Adapted and redrawn from J. W. Schopf, 1978, "The evolution of the earliest cells," *Scientific American* 239(3): 110–38.)

organelle that houses DNA) and are called **prokaryotes**. For about two billion years, all organisms were prokaryotes, but by about 1.5 BYA the fossil record shows organisms with a discrete nucleus. These are called **eukaryotes** and include all organisms, past and present, that are more advanced than the bacteria and cyanobacteria. The combination of high oxygen concentration and efficient systems for using it are the likely factors that led to the incorporation of DNA into a nucleus. Eukaryotic cells probably originated from the **symbiosis** (living together) of several kinds of prokaryotes. The **mitochondria** (cell organelles that are the site of energy reactions) of eukaryotes are thought to be descendants of early bacterial cells. Similarly, the **chloroplasts** (chlorophyll-containing organelles) of plant cells are vestiges of blue-green algae. These early eukaryotes, in turn, led to the explosive proliferation of multicellular organisms that appear in the fossil record at the beginning of the Cambrian Period, 590 million years ago (Figure 18). What we begin to see here is the retention and modification, by more advanced organisms, of much of what went before.

We do not know if the scenario set out above—proteinoid microspheres to prokaryotes to eukaryotes—is the actual pathway taken by life, *but we do know that such a pathway is possible,* and that it is entirely consistent with what is known about the early Earth, modern life, and biochemistry. We can be sure that, whatever the pathway, it would have been a stepwise one. New data, new discoveries, and new intuitions will most likely lead to a revision of this scenario.

Emergent Evolution

Are Life and Non-life Really So Different?

The derivation of life from nonliving matter (**emergent evolution**), as just described, should not be lumped with the sort of **spontaneous generation** of complex organisms disproved so elegantly by Louis Pasteur. Emergent evolution deals with the stepwise origin of living things from non-living matter through numerous increases in chemical complexity. This has been shown to be a highly probable result of a real-world process very like the various simulation experiments that scientists have conducted. Creationists ignore this reality in their efforts to portray emergent evolution in the same light as the alleged origin of mice from dirty clothes, the appearance of maggots in rotting meat, or the growth of bacteria in soup, the nineteenth-century suppositions disproved by Pasteur.

Is new life still being evolved from nonliving components today? Unlikely, but because of the nonrandom, self-ordering properties of amino acids, the new life may be indistinguishable from the old life. It is also likely that newly arisen protocells (if emergent evolution gets this far today) would not be able to compete with modern organisms, or would be quickly gobbled up by them. It may also be that today's oxygen-rich atmosphere is not as conducive to the origin of prebiotic macromolecules.

The Origin of Multicellularity

Where Did Invertebrates Come from?

Many biologists consider that some single-celled, flagellated (whiptailed) eukaryote is ancestral to all the multicellular plants and animals. These unicellular organisms show some plantlike features (many are photosynthetic) and some animal-like features (they are highly motile, and lack cell walls). Multicellular organisms most likely arose through the aggregation of single-celled organisms. Many such colonial forms exist today. This arrangement has the advantages of increased division of labor, coordination of activity, and interdependence among cells. Both botanists and zoologists trace the origin of their subjects to hollow, spherical, colonial flagellates that resemble an embryonic stage of the more advanced organisms. The Plant and Animal Kingdoms are both thought to have developed from such a common ancestor.

Along the animal line, one of the earlier groups to evolve was the coelenterates (hydra, jellyfish, corals), which are postulated to have arisen via a larva-like stage called a planula, which is a tiny ciliated, free-swimming, pear-shaped mass of two cell types with no left or right side, and no head or tail end. Most animal groups beyond the coelenterates are **bilaterally symmetrical**—they have left and right sides and head and tail ends. These animals, called the Bilateria, have the advantage of concentrating the sense organs in a "head" region and are more or less streamlined for active movement. They are thought to be derivatives of a planuloid ancestor that eventually gave rise to the flatworms and, ultimately, to the great diversity of the other invertebrate groups.

A major division in the Animal Kingdom occurred soon after the development of bilateral symmetry. One of the two lines that followed from that division led to the animal group in which the blastopore (the external opening of the gut) of an embryo develops into a mouth. These animals, which include various types of worms, mollusks, and arthropods, are called Protostomia. The other line led to the group whose blastopore becomes an anus. These animals, called Deuterostomia, consist of the echinoderms (sea stars and their relatives), hemichordates (some marine worms), and the chordates (tunicates, amphioxus, and vertebrates, including humans).

The echinoderms and the hemichordates have very similar ciliated larvae, and they probably share a common ancestor. Though the hemichordates and chordates share certain fundamental features, the hemichordates, until recently classified with the chordates, are now considered a separate group between the echinoderms and chordates. The prevailing view in zoology today is that both the echinoderms and the chordates evolved from a common ancestor in the remote past. The evidence for this view rests chiefly on their similar embryonic organization and development, not their adult features.

The Origin of Vertebrates
Where Did Backbones Come from?

One should not suppose that the evolution of the great variety of lifeforms on Earth proceeded in a straight line from primitive cells to humans. That simple notion is the old-fashioned scale-of-nature concept made obsolete by Darwin. Evolution proceeds like a growing bush, with branches radiating in various directions. There is insufficient space here to elaborate on the evolution of all of the many animal or plant groups, but I will recount some of the most interesting pathways along the branches that gave rise to our species.

The following is the most widely accepted scenario, but it is certainly open to such revision as might be suggested by new information. Vertebrates may have arisen from a protochordate similar to tunicates (sea squirts). Adult sea squirts are unlikely-looking vertebrate ancestors. They are small, baglike animals that spend their lives attached to the sea floor, but they have a larval stage that resembles a tadpole (Figure 19). This free-swimming larva has the characteristic features of a chordate, such as a dorsal hollow nerve cord, a **notochord** (a stiff rod of cells below the nerve cord that serves as internal support), and gill slits (perforated openings in the pharynx that evolved as a filter-feeding apparatus and led to the evolution of internal gills; in the higher vertebrates—reptiles, birds, and mammals—all traces of the gill slits usually disappear in the early embryonic stages). It is hypothesized that some tunicate tadpole larvae attained sexual maturity, a process called **neoteny** (achievement of sexual maturity at an embryonic or juvenile stage). This development opened the door for a great deal of **adaptive radiation** (the tendency of successful forms to spread out into all available ecological niches) into many new forms. Continued evolution of the neotenous larvae may have led to the jawless ostracoderms, the first true vertebrates, which appeared in the late Cambrian Period, over 500 MYA. It is from the small, ancient, armored ostracoderms that the various fish groups derived.

Today's lampreys are descendants of the ostracoderms. They are eel-like animals that retain the jawless condition with a lack of paired fins, but they have lost their armor plating. Jawed fishes evolved not from the lampreys but from a different group of ostracoderms. These new fishes had—in addition to jaws—paired fins and bony armor. They are known as placoderms. Their jaws developed from modifications of the first two gill arches (the supports for the gills), which eventually curved into the typical

Figure 19. Possible scenario for the origin of vertebrates. Some of the free-swimming tadpole-like larvae of ascidians (tunicates or sea squirts) acquired sexual maturity while still in the larval stage. Over some 500 million years, this hypothetical vertebrate ancestor gave rise to the first vertebrates, jawless fishes called ostracoderms, and the ostracoderms in turn gave rise to the first jawed fishes, the placoderms. (Adapted and redrawn from C. P. Hickman Jr., L. S. Roberts, and F. M. Hickman, 8th ed., 1988, *Integrated Principles of Zoology*, by permission from C. V. Mosby Co., St. Louis.)

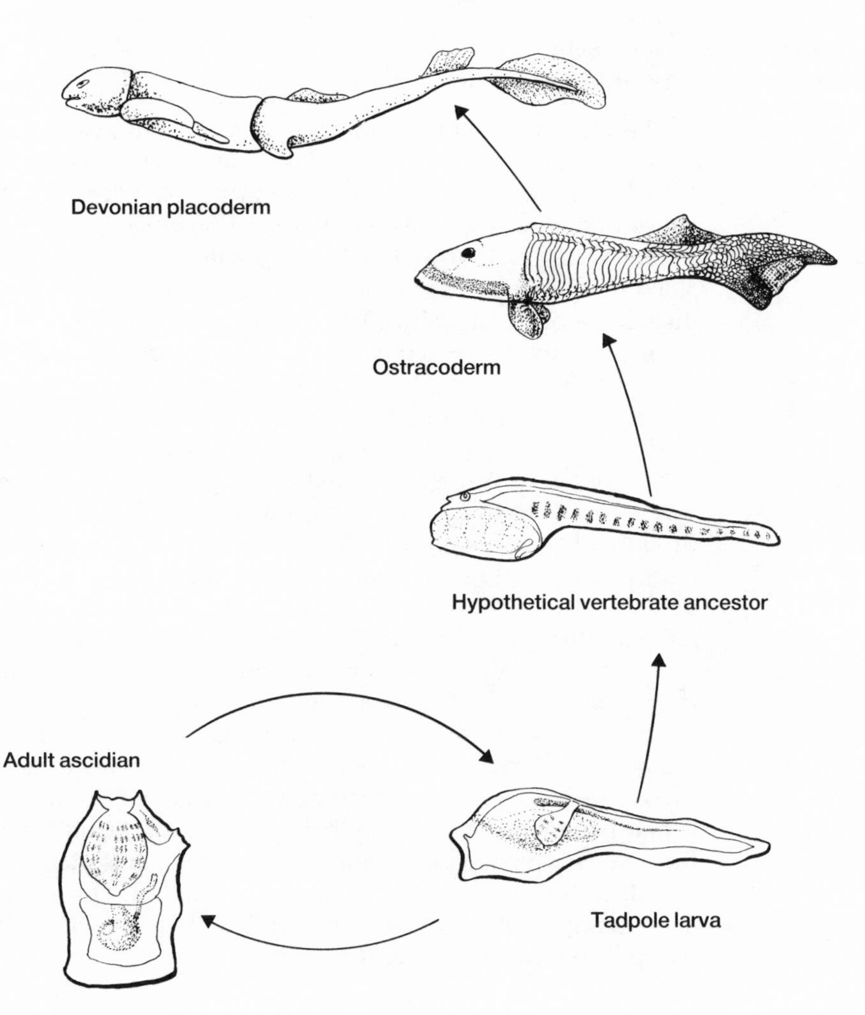

position of vertebrate jaws. Teeth developed on these surfaces as a derivative of skin. Paired fins formed from lateral folds in the body wall. The fins and jaws combined to allow active predation, and opened up new food sources that made possible an entirely new life style. The placoderms gave rise to the bony fishes and to the sharks before vanishing in the Devonian.

The Terrestrial Vertebrates
What Led Them to Crawl up on Land?

The Devonian Period, ensuing 400 MYA, witnessed a major step in the history of life. At this time, a primitive group of bony fishes with lobe-fins, called rhipidistians, flourished. These fishes had muscular limbs built around bony skeletons that resemble those of four-footed land animals. Such structures enabled them to move about on the bottom of swamps. The rhipidistians, though basically gill breathers, were capable of gulping air and extracting oxygen from it via their lungs. (In other fishes the lungs became modified as a swim bladder, or buoyancy organ). The fossil record indicates that the Devonian was a time of droughts, and the rhipidistians, with their ability to utilize atmospheric oxygen and to shuffle over land, were able to survive by moving from one drying water hole to another. As the droughts became more severe, natural selection favored those animals that could survive out of water. It is ironic, but perhaps inevitable, that the colonization of land was accomplished by animals attempting to retain the aquatic environment.

A second group of lobe-finned fishes related to the rhipidistians flourished in the Mesozoic, then all but died out. One species, the coelacanth, *Latimeria chalumnae* (Figure 20), still exists today, in the Indian Ocean off the coast of Africa near Madagascar. Its discovery, in 1938, electrified ichthyologists, who thought it had been extinct for 70 million years. It too bears the lobed fins and is considered a "living fossil," by which is meant that it manifests the phenomenon of arrested evolution. The cases of 34 such living fossils have been compiled by Eldredge and Stanley (1984; listed under Further Reading).

Figure 20. The coelacanth, *Latimeria chalumnae,* a lobe-finned crossopterygian fish once thought to have been extinct for 70 million years. The first known living specimen was discovered off East London, South Africa, in 1938. Since then at least 130 others have been caught off the Comoro Islands in the western Indian Ocean, between Africa and Madagascar, 1,900 miles northeast from where the first specimen was caught. The coelacanth lives in depths of 500–1,300 feet off steep rocky shores and moves into shallower waters in search of prey at night. It uses its remarkable fins to stabilize its body while drifting in offshore currents. This species, which gives birth to live young, can reach nearly 6 feet and weigh 180 pounds. The fossil record of lobe-finned fishes numbers about 30 species and dates back 400 million years, into the Devonian (see Figure 21). This group of fishes is clearly similar to the ancestor of all land vertebrates, and the discovery of a living member of the group is regarded as one of the most important biological and evolutionary finds of the century. The describer of this species, the South African ichthyologist J. L. B. Smith, called coelacanths "machines for reading time backwards." (Drawing by D. P. Voorvelt, courtesy of the J. L. B. Smith Institute of Ichthyology.)

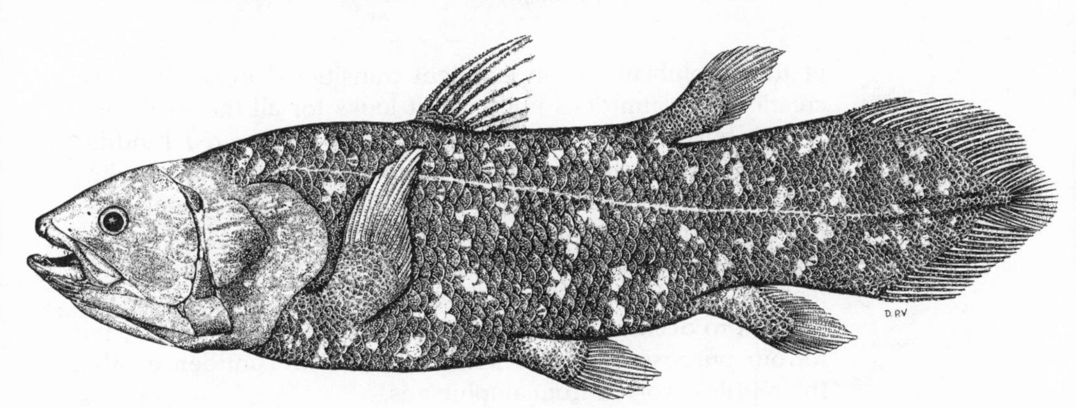

The skulls of rhipidistian fishes are remarkably similar to those of the early amphibians, the first group of land vertebrates. Both have internal nostrils and peculiar teeth in which the enamel forms ridges that extend into the deeper layers of the tooth. This type of dentition is called labyrinthodont, and it is found only in rhipidistians and primitive amphibians. *Eusthenopteron* (Figure 21), a likely fossil candidate for the ancestor

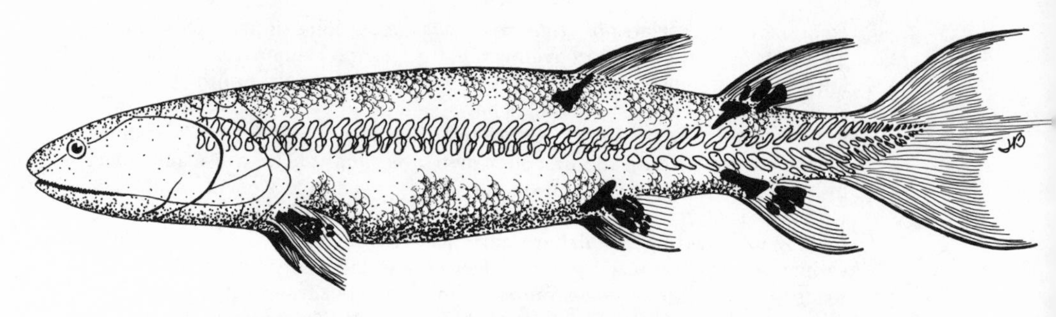

Figure 21. Eusthenopteron foordi, a lobe-finned crossopterygian fish of the Devonian period. This fish, over 3 feet long, was near the line of ancestry of the early land-dwelling amphibians. Note the muscular fins and other skeletal elements homologous to the structures of terrestrial vertebrates. (See Figure 3.) (Adapted from a drawing in S. E. Luria, S. J. Gould, and S. Singer, 1981, *A View of Life*, Benjamin/Cummings Publ. Co., Menlo Park, Calif., and used by permission.)

of the amphibians, is an excellent transitional fossil, the type creationists claim does not exist; it looks for all the world like a fish, but its paired forefins (pectorals) and paired hindfins (pelvics) all contain the bony structure needed by land-crawling vertebrates. Another excellent example of a transitional fossil is a semi-aquatic animal called *Seymouria* (Figure 22), from the lower Permian of Texas. It shows both amphibian and reptilian features. The mixture of characteristics is so complete that it is difficult to decide *Seymouria*'s proper taxonomic placement, but for our purposes it is sufficient to note—with confidence—that the reptiles evolved from amphibians.

The major contribution of reptiles was the development of the **amniotic egg** (shelled egg), which protected the developing embryo in a fluid-filled sac and freed the reptiles from reproduction in water. Their eggs were self-contained and could be laid on land in protected situations. This gave the early reptiles a great advantage over the amphibians, whose eggs and young had to pursue life in the water, and no doubt contributed to the decline of the amphibians. (We have already discussed the origin of birds from reptiles via *Archaeopteryx*; see Figures 10 and 11.)

The first mammals, which derived from the mammal-like reptiles called therapsids (Figure 23), appeared in the late Triassic, about 213 MYA. Following the extinction of the dinosaurs and other ruling reptiles at the end of the Cretaceous, 65 MYA, the relatively small and inconspicuous early mammals, which had been getting along all that time at a modest level, radiated into a variety of forms, making the most of their nursing and protection of young, their warm, constant-temperature blood, and their live births. One of these early mammalian forms was the ancestor of the primates.

Figure 22. Reconstruction of *Seymouria*, a Permian fossil classified as an amphibian but showing a mix of amphibian and reptilian skeletal characteristics. The skull has many features in common with labyrinthodont amphibians, and the postcranial skeleton shows derived characters of the early reptiles.

Figure 23. Reconstruction of *Cynognathus*, a lower Triassic reptile (therapsid) about the size of a wolf that approached the mammalian form.

Primate Relationships
Who Is Related to Whom?

The ancestor of the primates, which today include lemurs, monkeys, apes, and humans, was most likely a small, insect-eating mammal similar to the squirrel-like, modern-day tree shrew, *Tupaia*, of the Orient. Primates as a group have five

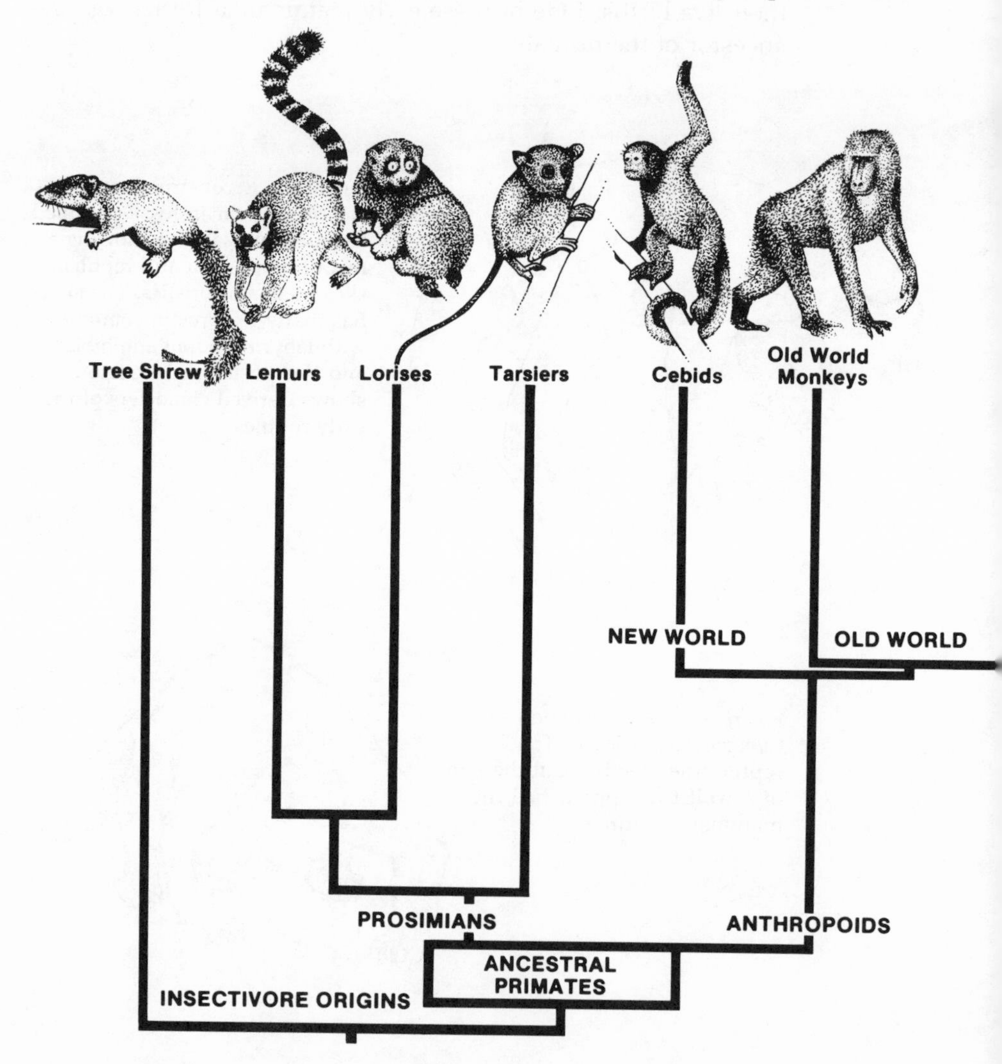

digits on each limb (usually with nails instead of claws), opposable thumbs and/or big toes, four types of teeth, binocular vision, a well-developed cerebrum (upper brain), and relatively slow postnatal development. Primatologists place the many primate species in two major groups (Figure 24). The prosimians are small, primitive, arboreal (tree-dwelling) primates such as lemurs, lorises, and tarsiers. The anthropoids, the more advanced

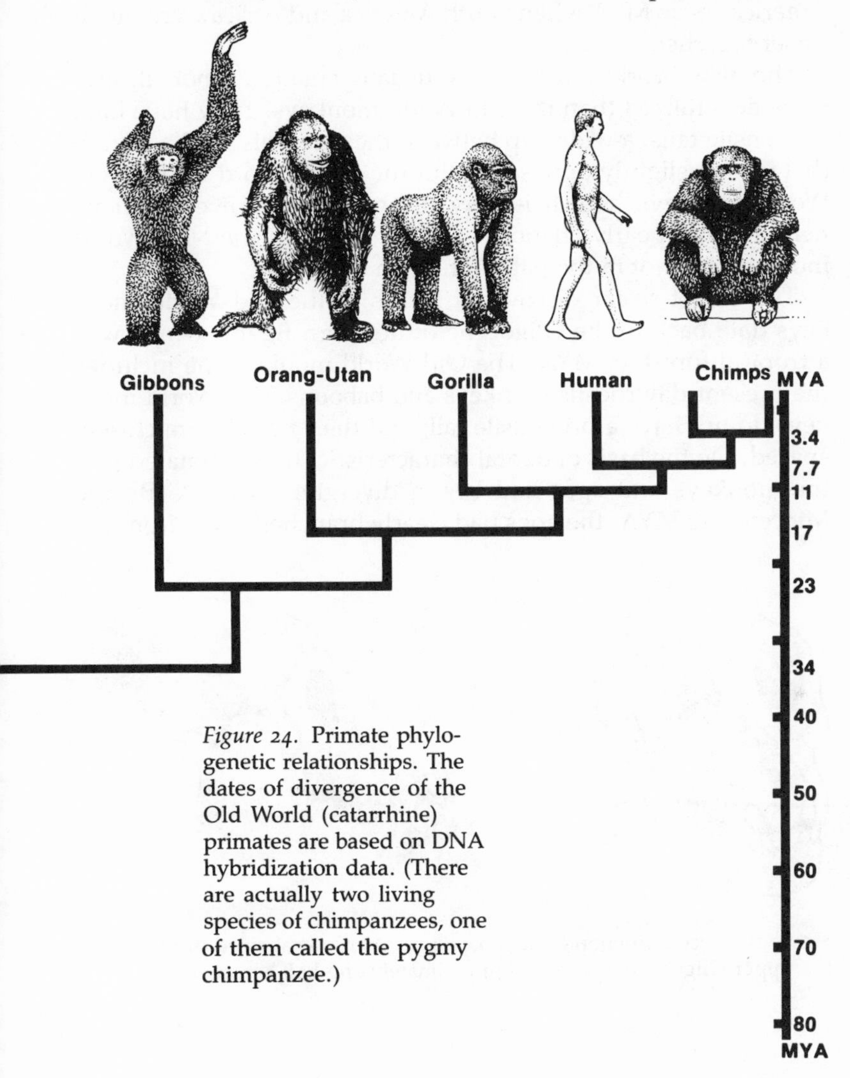

Figure 24. Primate phylogenetic relationships. The dates of divergence of the Old World (catarrhine) primates are based on DNA hybridization data. (There are actually two living species of chimpanzees, one of them called the pygmy chimpanzee.)

primates, include the New World monkeys, Old World monkeys (including baboons), apes (siamang, gibbons, orangutan, gorilla, chimpanzees), and humans.

The ancestors of the monkeys evolved roughly 60 MYA and split into two groups, the New World monkeys and the Old World monkeys. Recent speculation suggests that the New World monkeys arose from African ancestors that drifted to South America 35–40 MYA when South America and Africa were much closer together.

The New World monkeys are usually smaller, arboreal, and more generalized than the Old World monkeys. They have long prehensile tails, a wide gap between their nostrils, and a thumb that is only slightly opposable. But the fossil record of the New World monkeys, which is very meager, need not concern us here, for as an early offshoot of the prosimians, the New World monkeys are not in the path of human evolution.

The oldest so-far discovered fossils of the Old World monkeys date back to the Oligocene of northern Egypt, which was a tropical forest 38 MYA. The Old World monkey line includes the present-day rhesus monkeys and baboons. Old World monkeys do not have a prehensile tail, and their nostrils are closely spaced. On the basis of dental characteristics, it is estimated that the monkeys and apes had begun diverging 35 MYA. By the Miocene, 20 MYA, the apes had clearly branched away from the

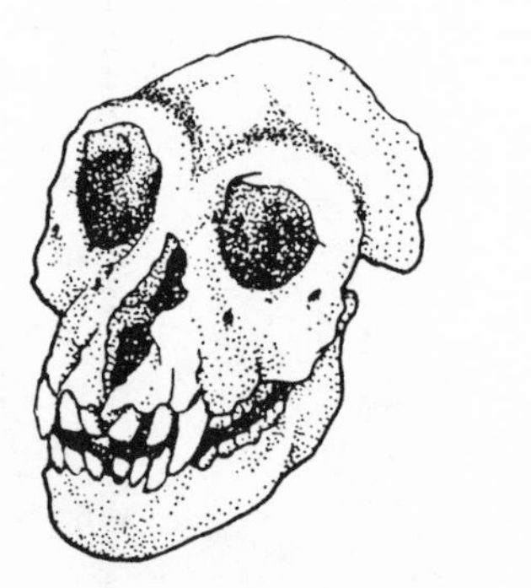

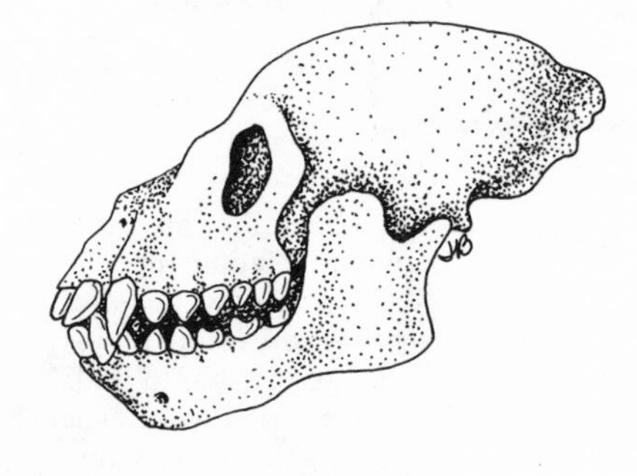

Figure 25. Reconstruction of *Aegyptopithecus zeuxis,* a cat-sized primate from the upper Oligocene of Egypt. Approximately one-half actual size.

Figure 26. Reconstruction of *Proconsul africanus,* a baboon-sized early Miocene ape possibly ancestral to all apes and humans. Approximately one-half actual size.

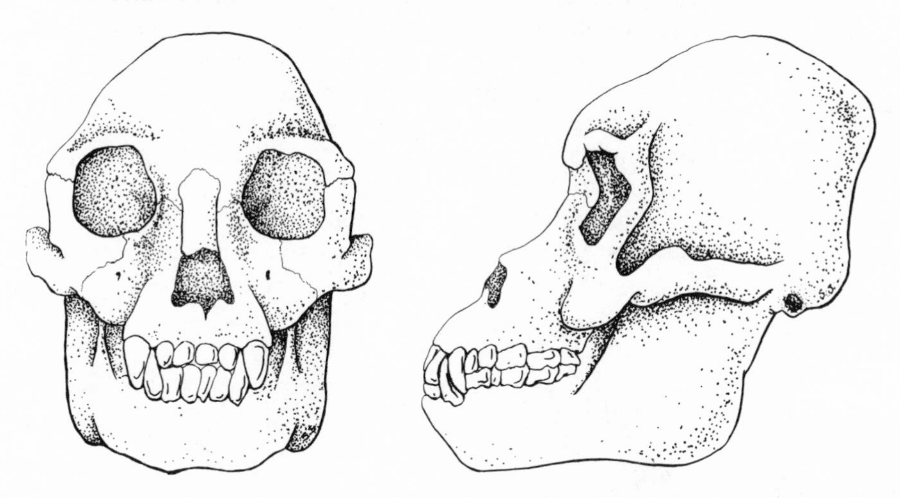

Old World monkeys and radiated into many different species.

Discoveries at higher (younger) levels in the Egyptian Oligocene have yielded the virtually complete skull and lower jaw of an ape named *Aegyptopithecus* (which means Egyptian ape; Figure 25). Its skull was monkey-like, but its jaws and teeth were like those of an ape. Recent finds in Egypt have tripled the supply of Egyptian ape skulls. These cat-sized fruit-eaters flourished about 32 MYA, and lived in social groups. They had relatively large brains for that time and substantial facial variation. The skulls of the newly discovered specimens are more like the African apes than the Asian species. Their discoverer, anthropologist Elwyn Simon, feels that the Egyptian ape may foreshadow an extinct African ape named *Proconsul* (Figure 26), which had monkey-like trunk, arms, and hands with an ape-like skull and teeth. *Proconsul* dates back to 20 MYA.

Apes, classified by most zoologists and anthropologists as the family Pongidae, are called pongids. The sister group of the apes is the family Hominidae, which contains the genera *Australopithe-*

cus (the so called "ape-men") and *Homo* (humans); as a group they are called **hominids**. When referring to pongids and hominids together as one larger category, anthropologists employ the term **hominoid**. There is a recent trend among some anthropologists to assign the African apes to the Hominidae to reflect the very close relationship between the apes and humans.

The Miocene was a time of tremendous adaptive radiation for hominoids. An abundance of hominoid fossil material has been found in the Old World from about 20 to 8 MYA. Unfortunately, interpretations of this material are as varied as the fossils themselves. Views and terminology have changed as more ma-

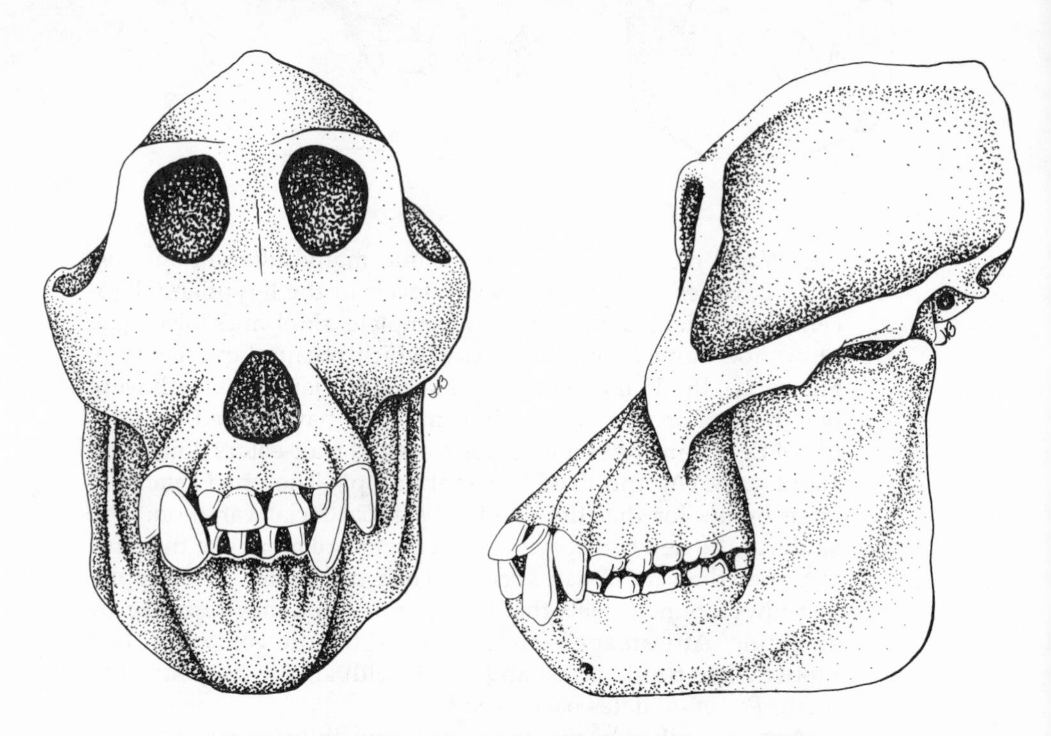

Figure 27. Reconstruction of *Sivapithecus indicus*, a large Miocene ape from Pakistan probably ancestral to the orangutan. Approximately one-half actual size.

terial has been discovered. There has also been a proliferation of scientific names, which makes discussion very difficult and confusing. The principal characters in this story are *Proconsul* from Africa, *Ramapithecus* from India, and *Sivapithecus* (Figure 27) from Africa and Asia. Early opinion placed *Ramapithecus*, which first appeared 12 to 14 MYA, as a hominid and a possible candidate for a human ancestor. This view is now out of favor with most anthropologists, who place the split of the ape-human lineage much more recently, perhaps 6 to 8 MYA. In fact, many workers in this field now consider that *Ramapithecus* should be united with *Sivapithecus*.

Recent finds of upper and lower jaw fragments, teeth, and pieces of femur (thigh bone) of *Sivapithecus* in Kenya by Alan Walker, a paleoanthropologist at Johns Hopkins University, and Richard Leakey date back to 17 MYA. Early speculation regarding those finds suggested that *Sivapithecus* gave rise to the common ancestor of hominids and the great apes in Africa. A branch of *Sivapithecus* that spread to Asia about 13 MYA gave rise to the orangutan. The African group then split into the three lineages that resulted in humans, chimpanzees, and gorillas. Alternatively, the new *Sivapithecus* material could just mean that the orangutan lineage is older than previously thought.

In jaw shape and enamel thickness, humans have retained the presumed primitive features of *Sivapithecus*; chimpanzees and gorillas have the more derived features of thinner enamel and different arrangement of jaw musculature. This agrees with the radical interpretation by science writers Gribbin and Cherfas, who postulated that chimps and gorillas may be derived from a humanlike ancestor, rather than the reverse. These speculations differ from the orthodox scenario that *Sivapithecus* was ancestral only to the orangutan and was outside the main line of human evolution. It is premature to evaluate conclusively the rapid developments in paleoanthropology. No consensus exists yet, among anthropologists, but it is the branching pattern of the hominoid tree, not its existence, that they debate. Time and more fossils will elaborate and clarify the picture.

The Molecular Clock and DNA Hybridization
Can We Actually Measure Ancestry?

When did the ape-human lineage split? So far, fossil discoveries have not yielded a clear-cut answer. There is a large gap in the fossil record of the hominoids from about 8 to 4 MYA, and we do not have fossils to indicate when the gorilla-chimpanzee-human lineage split from a common ancestor. Scientists have turned to biochemical methods to answer this question. Two such techniques, the **molecular clock** and **DNA-DNA hybridization**, will be discussed.

The principle behind the molecular clock is simple. Related species such as chimpanzees and humans share a common ancestor somewhere in their past evolutionary history. If neutral changes in their proteins (discussed in Chapter 1) have accumulated at a constant rate, the amount of molecular differentiation between related species reflects the time since their divergence. The difference in amino acid sequence of the blood protein albumin among humans, chimpanzees, and gorillas is only 1.2%.

What does information of this sort tell us? First, the fossil evidence indicates that the Old World monkeys diverged from the ape-human ancestor about 25–35 MYA. The index of dissimilarity between the albumin of Old World monkeys and that of the ape-human lineage is 2.3 units, which corresponds to roughly 30 million years. Other studies have shown that a nonlinear but relatively simple formula identifies the relationship between time and distance. The index of dissimilarity between humans and chimps has been found to be just 1.17 units. If this value is put into the formula, the date of divergence turns out to be about 5 MYA. Humans and apes are genetically very similar. Much of their difference is probably due to a very few genes called **regulator genes**, which control the expression of a few structural genes, which in turn affect the relative growth of different body parts. Thus humans and apes are not as genetically different as they look.

Some **paleoanthropologists**, however, have been reluctant to accept molecular-clock evidence. They suspect an earlier diver-

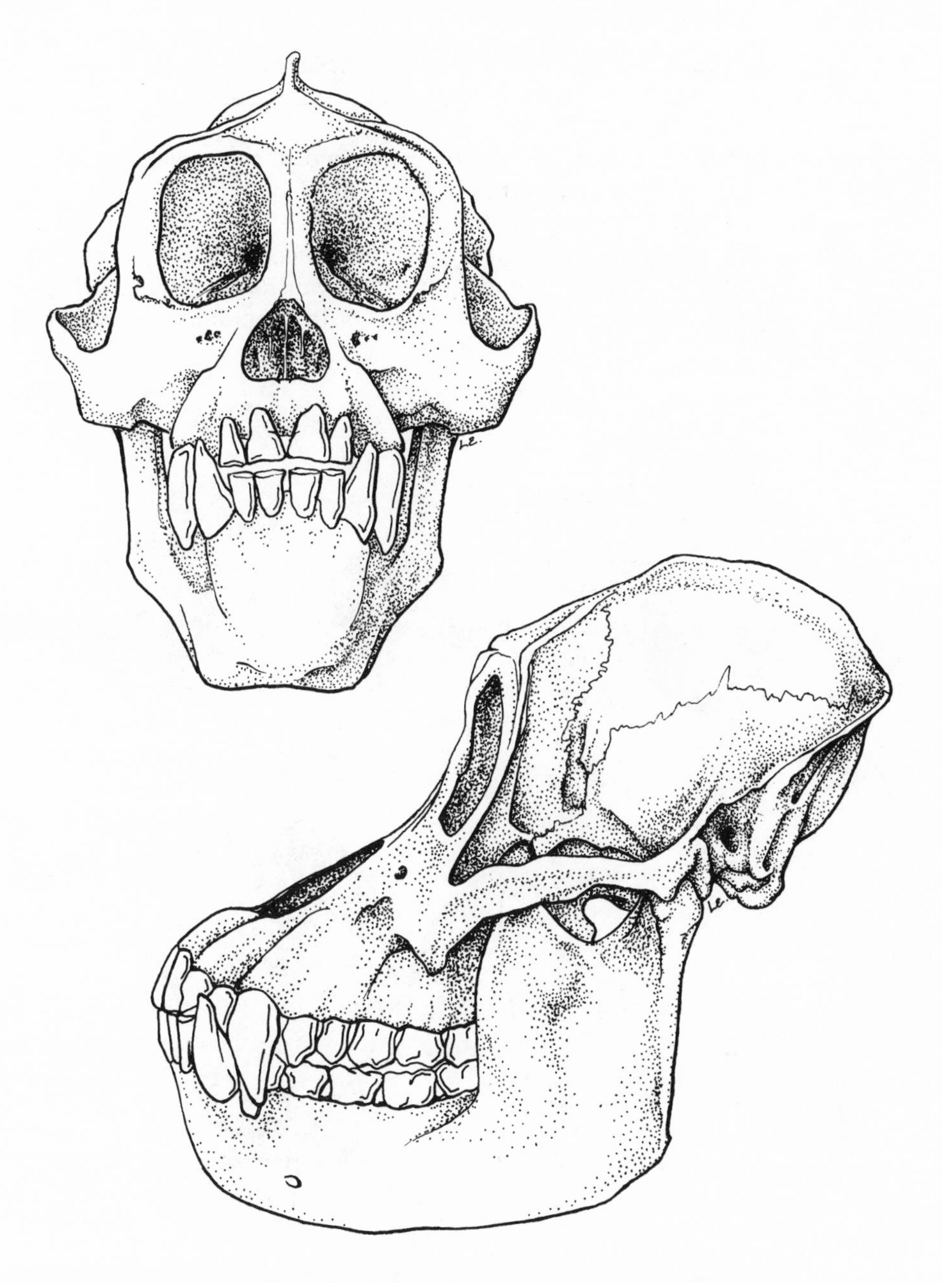

Figure 28. Orangutan, *Pongo pygmaeus*, male. One-half actual size.

gence date, and will not be convinced of the molecular dating until more fossil material becomes available.

The single-protein studies mentioned above offer a reasonably good figure for divergence, but each protein evolves at its own rate and represents only a tiny portion (a few hundred to a few thousand nucleotide bases) of the **genome** (the totality of

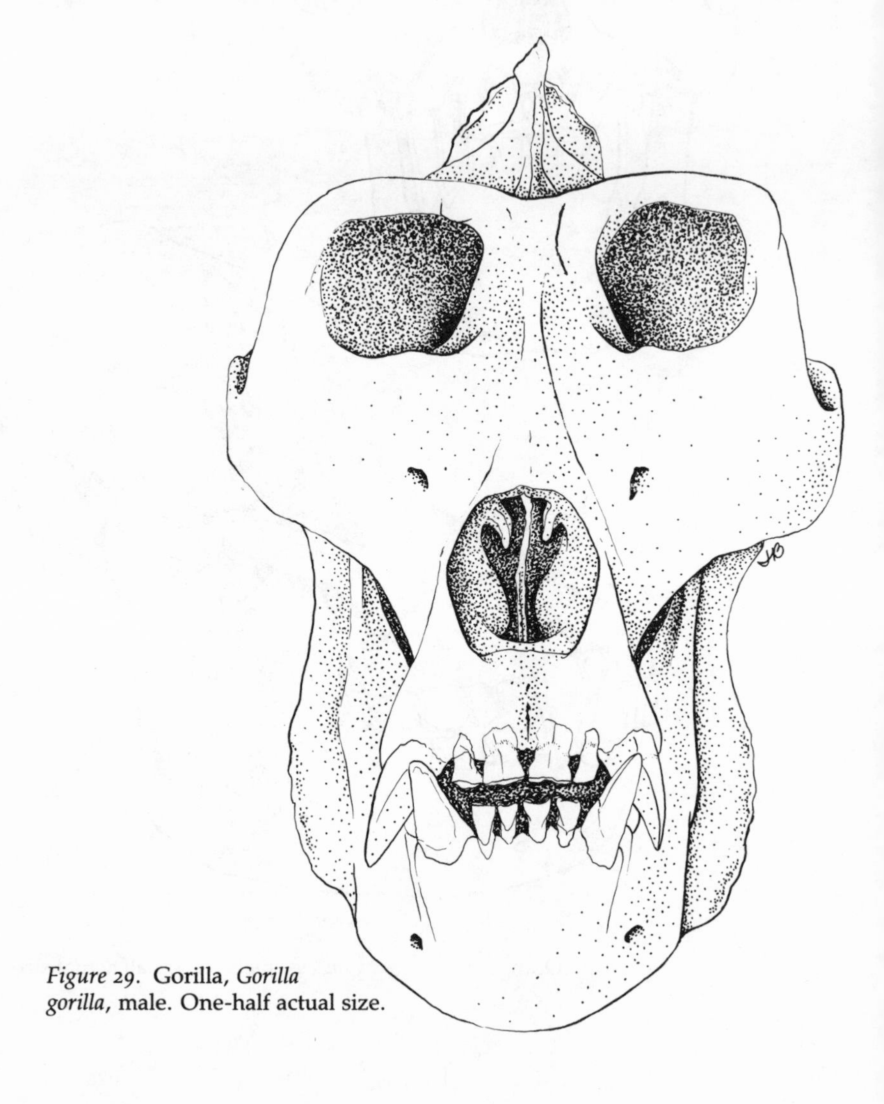

Figure 29. Gorilla, *Gorilla gorilla*, male. One-half actual size.

genetic information in an organism). Thus protein clocks tend to be sloppy. Data from protein A may give a divergence date slightly different from that yielded by protein B. The data from DNA-DNA hybridization, on the other hand, are thought to keep excellent time, because they average billions of nucleotides over millions of years. They deal with the entire genome, which

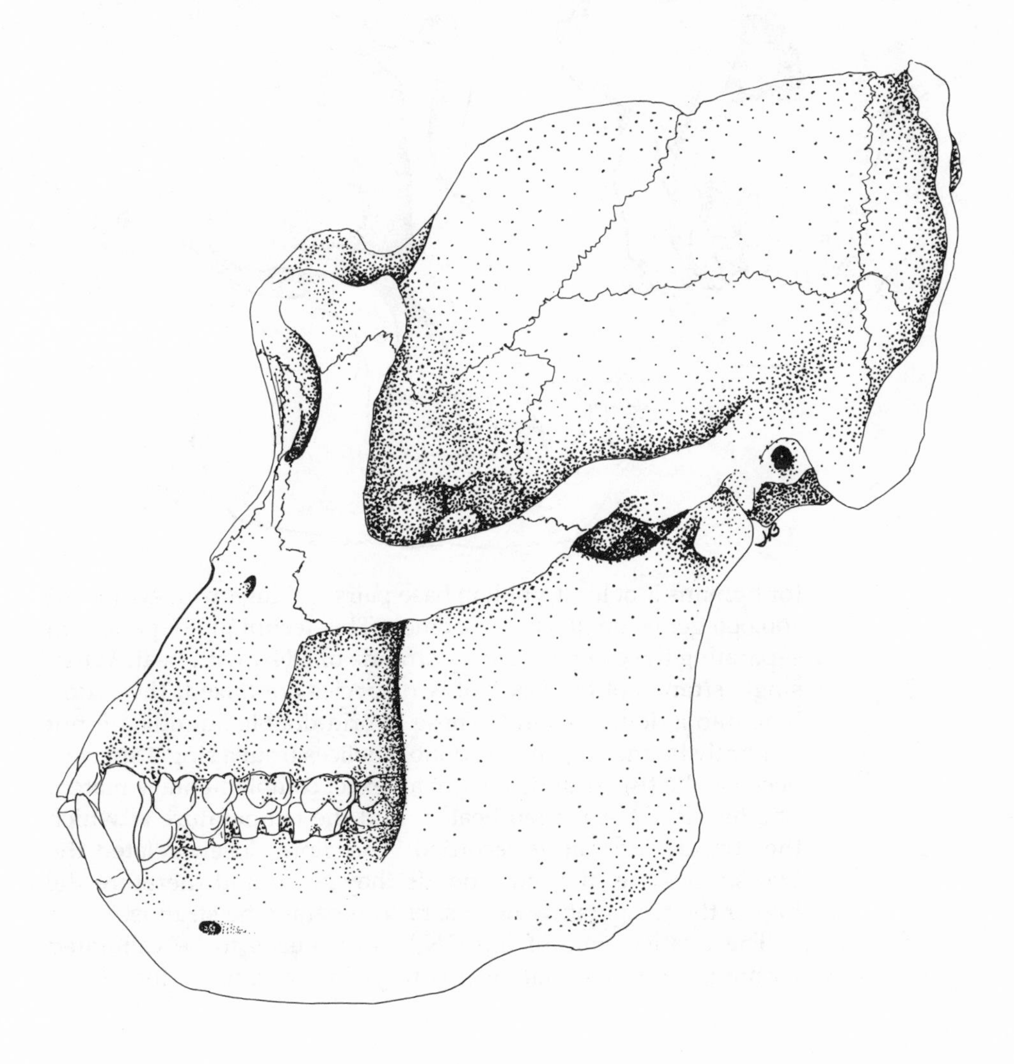

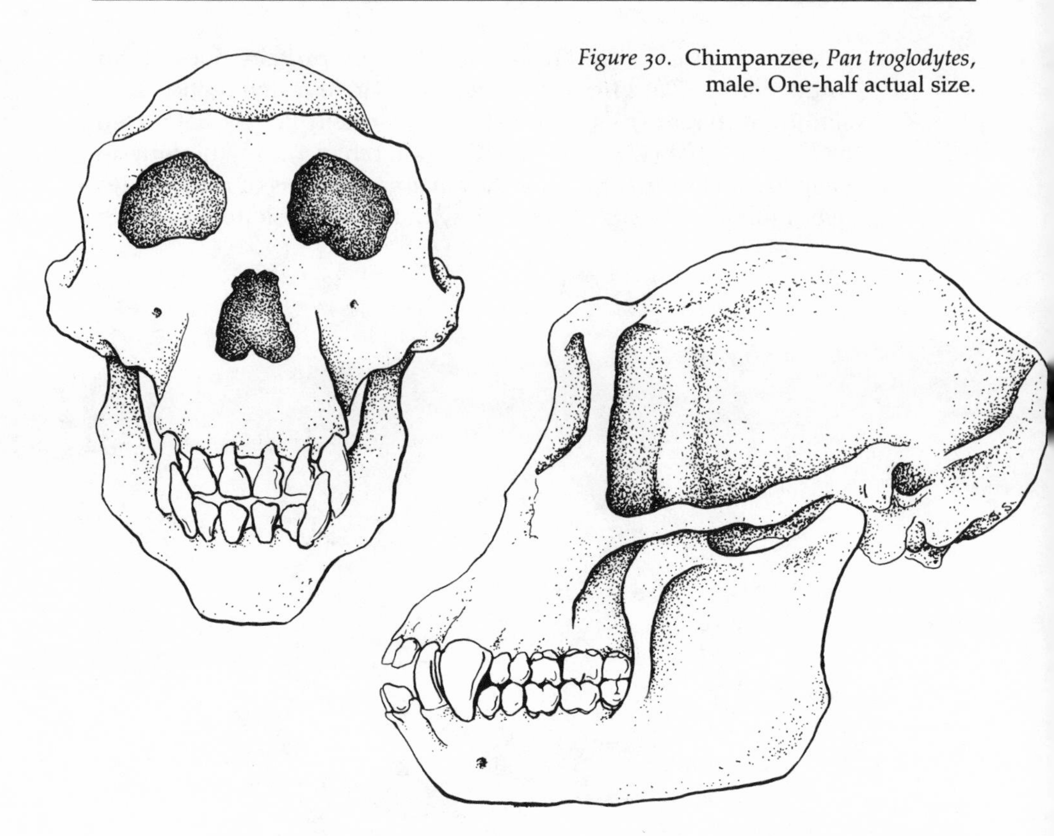

Figure 30. Chimpanzee, *Pan troglodytes,* male. One-half actual size.

for humans is at least 3 billion base pairs making up an estimated 100,000 genes on 46 chromosomes. This technique depends on separating the complementary strands of DNA with heat. When single strands of the DNA of two species are mixed, they combine into a double strand. These hybrid double strands are not as tightly bound together as double helices from a single species, because the two strands are not an exact complementary match. The hybrid DNA is then heated, and the temperature at which the strands separate is recorded. The more closely related the two species are, the more bonds they share and therefore the higher the temperature necessary to separate the strands.

The relative time of the DNA clock needs to be calibrated with a secure fossil date if it is to yield the time of divergence accurately. The most recent study of Sibley and Ahlquist (1987) provided a minimum and a maximum divergence date for each hominoid species. The minimum date is tied to the fossil dating of the Old World monkey lineage at 25 MYA, and the maximum date is based on the fossil dating of the orangutan branch at 17 MYA. Using these calibration points to set the DNA hybridization clock, the dates of divergence of the various primate groups (Figure 24) are as follows: Old World monkeys (Cercopithecoidea), 25–34 MYA from the apes and humans; gibbons (*Hylobates*), 16.4–23 MYA from the other apes and humans; orangutan (*Pongo*; Figure 28), 12.2–17 MYA from the African apes and humans; gorilla (Figure 29), 7.7–11 MYA from the chimpanzees and humans; and humans, 5.5–7.7 MYA from chimpanzees (*Pan*; Figure 30). *These findings indicate that humans and chimps are more closely related to one another than either is to the gorilla*. The two chimp species, common and pygmy (*Pan troglodytes* and *P. panis-*

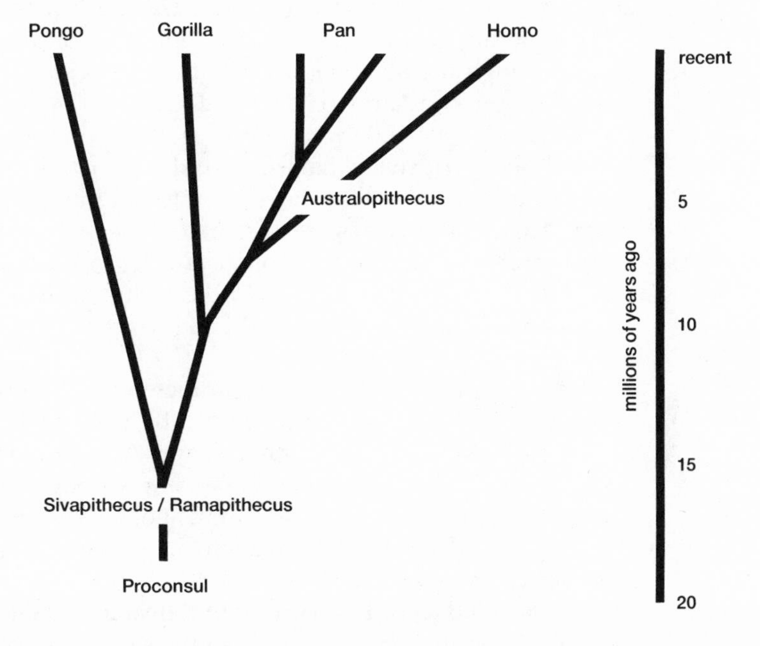

Figure 31. Hominoid evolution, as determined from fossil and DNA hybridization dates.

cus) diverged 2.4–3.4 MYA. It is this scheme that is reflected in Figures 24 and 31. If the date of the calibration fossils shows that the orangutan lineage diverged, say, 13 MYA, instead of 17 MYA, then the maximum divergence dates given above will need to be multiplied by a factor of about 0.76 to reflect a later separation.

Scientists routinely check the findings of their colleagues, by repeating their studies, especially when the importance of the results is great. In this case, Caccone and Powell (1989) repeated Sibley and Ahlquist's work, using slightly different but more precise methods, *and found exactly the same results*.

The Australopithecines

How Far from Lucy to Us?

The australopithecines, which emerged roughly 4 MYA, are sometimes called "ape-men" or "missing links" because in many characteristics they are intermediate or transitional between apes and our genus, *Homo*. Evidence from endocranial casts indicates that, although the australopithecine brain was small (about the size of an ape's), it already exhibited signs of remodeling along the path of *Homo*. Likewise, fossil skulls, pelvises, and knee joints show conclusively that the australopithecines were upright walkers, but no stone tools have been found in association with them. Australopithecines are classified in the family Hominidae, and, therefore, many anthropologists refer to them as the first humans. Others prefer to reserve the term "human" for *Homo erectus* and *H. sapiens*, which had much larger brains and made stone tools.

The gap in the fossil record of primate evolution that begins about 8 MYA has been shortened by the well-publicized work of Donald Johanson and his discovery of "Lucy" (40 percent of a complete skeleton—actually 80 percent is known, because of bilateral symmetry), which dates from about 3.5 MYA. Other authorities have dated Lucy from 2.9 to 3.2 MYA, but more recent discoveries have pushed the date of this species back to 4 MYA. Two molars and a jaw fragment from Kenya may date as far back as 5 MYA, but they may not be from the same species.

Lucy (Figure 32) and 13 other individuals of the same species

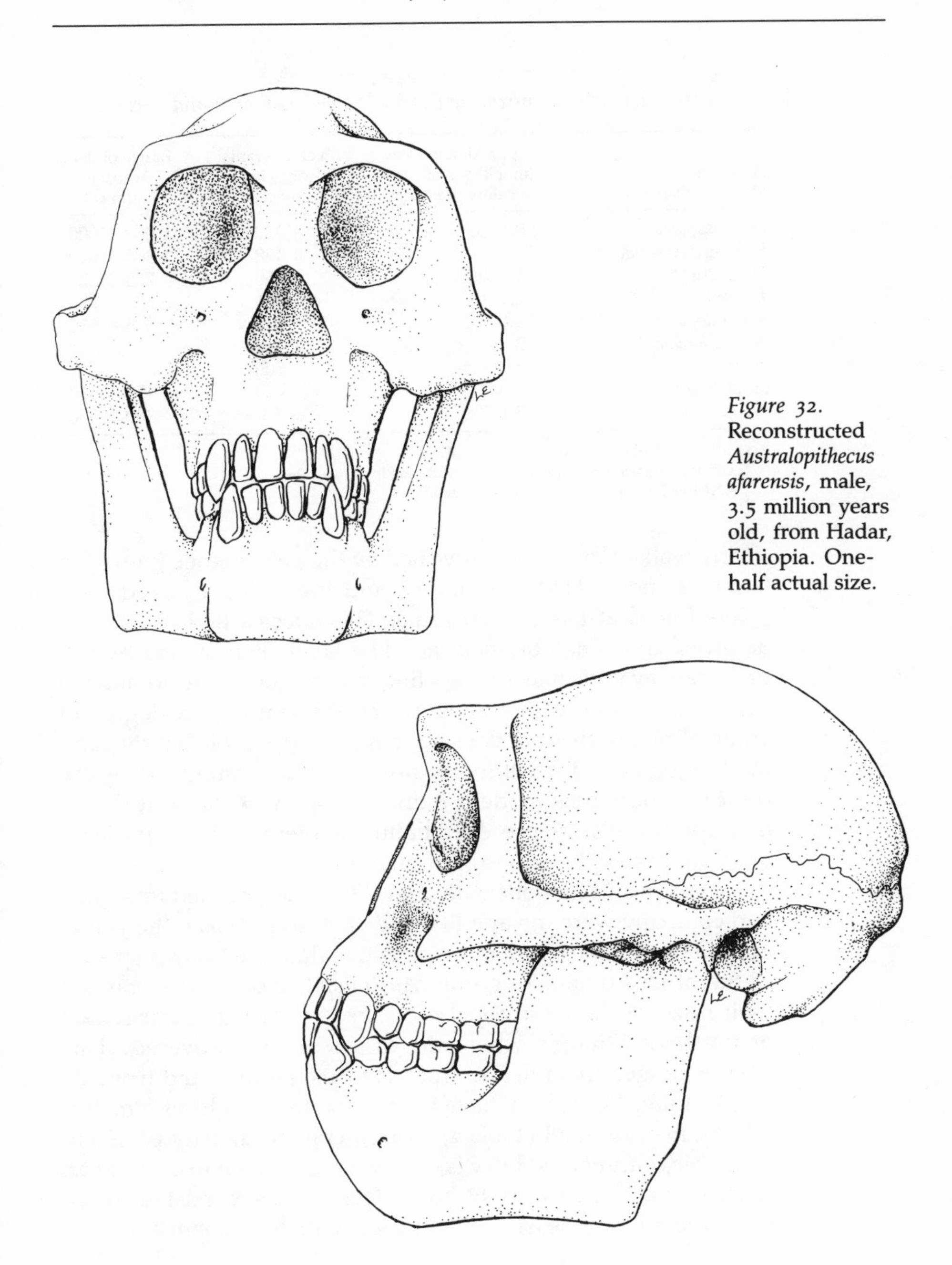

Figure 32. Reconstructed *Australopithecus afarensis*, male, 3.5 million years old, from Hadar, Ethiopia. One-half actual size.

TABLE 2
Brain Sizes and Approximate Dates of Some Fossil Hominid Species

Hominid species/subspecies	Approximate dates (in millions of years before present)	Average cranial capacity (in cc)	Range of cranial capacity (in cc)
Homo sapiens	0.5–present	1,330	1,000–2,000
H. s. neanderthalensis	0.15–0.03	1,438	1,300–1,610*
H. erectus	1.6–0.25	950	775–1,225
H. habilis	2.1–1.5	666	600– 752
Australopithecus boisei	2.0–1.2	515	506– 530
*A. aethiopicus***	2.5	410	
A. robustus	2.0–1.5	500	
A. africanus	2.8–1.9	450	435– 530
A. afarensis	3.8–2.8	440	380– 550

SOURCE: Campbell (1985).
*Data for "classic" Neanderthal from M. Shackley (1980).
**KNM-WT 17000. See Walker et al. (1986).

clearly walked upright, as revealed by the pelvis, knee joint, and femur found at Hadar, Ethiopia, and the well-preserved footprints found at Laetoli, Tanzania. This species had humanlike dentition in a small-brained, ape-like head. The fingers (which have narrow pads like an ape's fingers) and toes were somewhat curved like those of the apes, which may indicate a degree of arboreality, but the foot does not have an opposable big toe, and the toe joints and fossil footprints reflect the same type of upright locomotion as modern humans employ. Because it shows that upright posture preceded brain enlargement, this species is important to explanations of human origins.

Johanson and his colleague, Tim White, determined that Lucy and company were too ape-like to be placed in *Homo*, the genus of modern humans, and more primitive than the known species of the other humanlike group, *Australopithecus*. So they named their new species *Australopithecus afarensis*, after the Afar region of northern Ethiopia where the fossils were discovered. This species ranged in adult size from 55 to 150 pounds and from 3½ to 5 feet tall. Males may have been twice the weight of females. The brain was small (Table 2), with a cranial capacity of 380 to 500 cubic centimeters. Lucy is clearly a transitional fossil with an apelike head on an upright body. This species persisted in the fossil record for at least 1 million years with little change.

Figure 33. Reconstructed *Australopithecus africanus*, about 2.5 million years old, from Sterkfontein, South Africa. Originally named *Plesianthropus transvaalensis* and called "Mrs. Ples" by the media. Approximately one-half actual size.

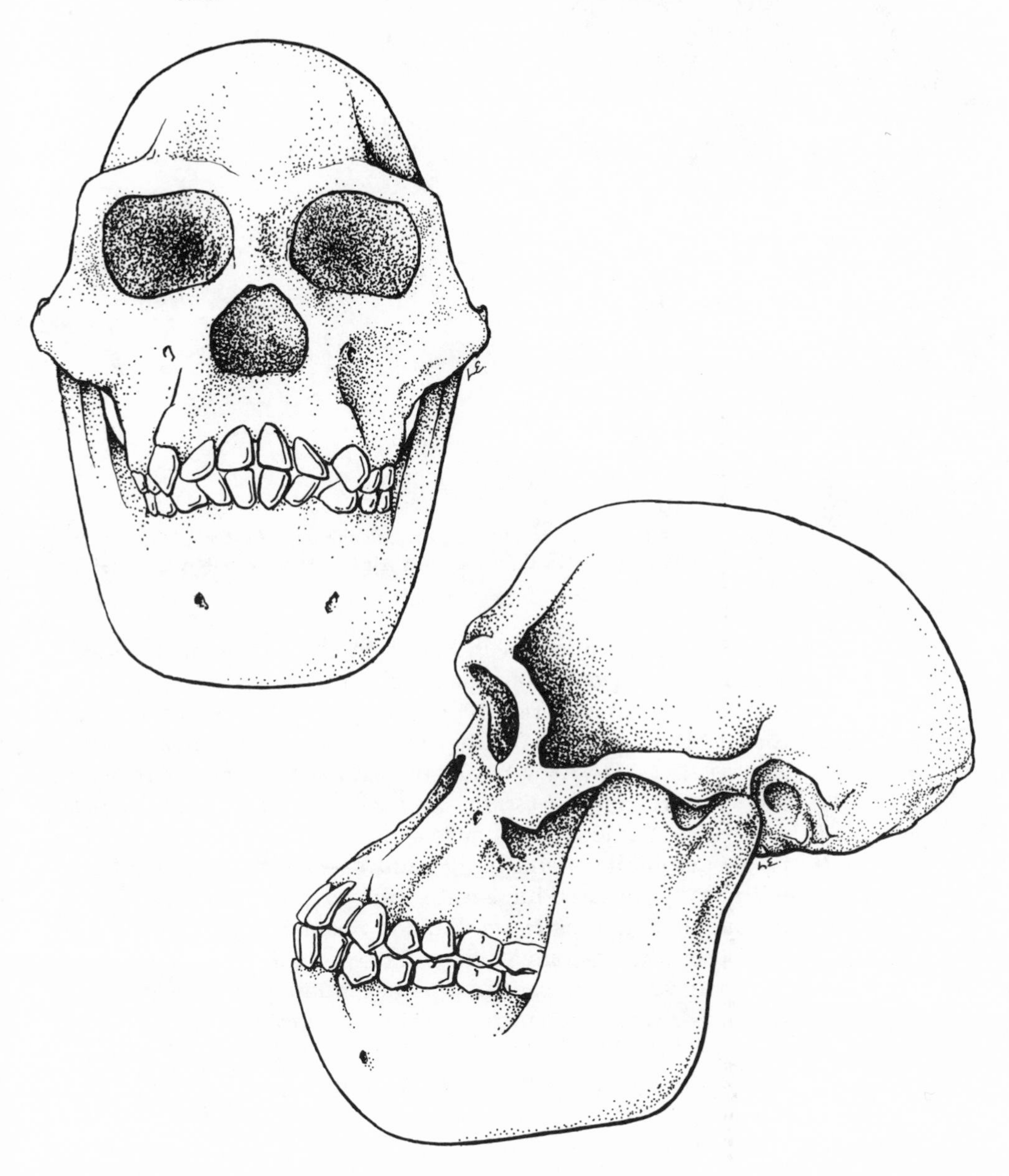

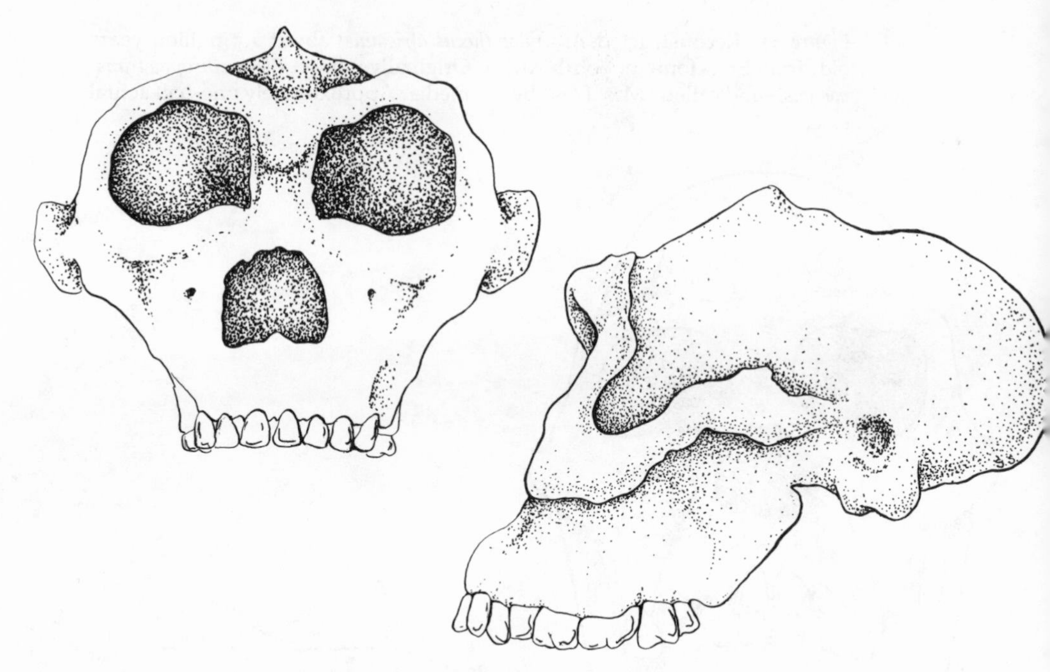

Figure 34. Partial reconstruction of *Australopithecus robustus,* about 1.7 million years old, from Swartkrans, South Africa. Approximately one-half actual size.

Until the discovery of the Black Skull in 1986, an emerging consensus forged by Johanson and others went something like this. *A. afarensis* gave rise to the gracile (slender) *A. africanus* (2.8 to 1.9 MYA, Figure 33), from which derived the heavily built *A. robustus* (2.0 to 1.5 MYA, Figure 34) from southern Africa and the super-robust *A. boisei* (2.0 to 1.2 MYA, Figure 35) from eastern Africa. Some workers have considered *robustus* and *boisei* to be geographical variants of the same species; these robust forms with heavy jaws and huge teeth were adapted for coarser, more fibrous diets than is *Homo*. Randall Susman (1988) recently reported on the anatomy of finger bones of 1.8 MYA *A. robustus,* which have broad, padded tips like humans. These pads are probably related to an increased blood supply and sensory nerve endings that allow delicate manipulation of objects. Susman suggested that this precision grasping would allow the use of such

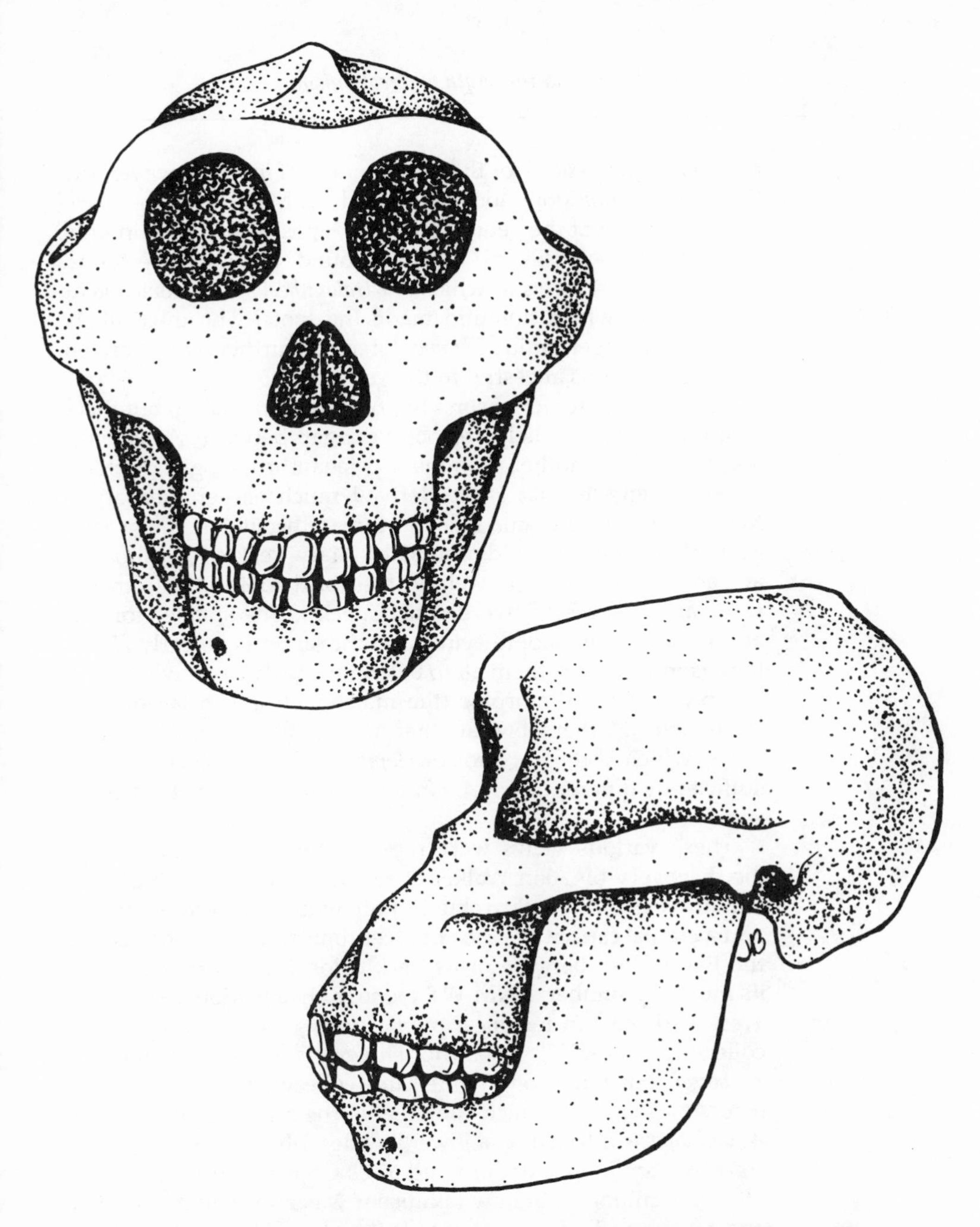

Figure 35. Reconstructed *Australopithecus boisei*, large male, about 1.8 million years old, from Olduvai Gorge, Tanzania. Originally named *Zinjanthropus boisei*, "Zinj" or "Nutcracker Man." Approximately one-half actual size.

tools as digging sticks for the procurement of food. However, the first evidence of stone tools is around 2.5 MYA, near the earliest appearance of the genus *Homo*. In deposits that contain only *Australopithecus* fossils, no tools are found, but tools are found in higher (later) levels in which *Australopithecus* and *Homo* occur together. So which hominid made the tools? The most likely guess is the larger-brained *Homo*. Johanson further hypothesized that *A. afarensis* is ancestral to the genus *Homo*.

The evidence for Johanson's hypothesis is appealing, but there is still an approximately 700,000-year gap between *A. afarensis* and *H. habilis*. Another prominent paleoanthropologist, Richard Leakey, suggests that *Homo* derived much earlier, perhaps 6 MYA, from an ancestor like the ramapithecines. Leakey suggests that fossil hominids found by Mary Leakey, his mother, at Laetoli, Tanzania, as well as fossilized footprints from the same area, represent two distinct species. The smaller form is an unknown australopithecine, and the larger is an early *Homo*. Johanson considers them all to belong to *A. afarensis*, which was notably sexually dimorphic (the males were much larger than the females). Other views are that *A. afarensis* gave rise to *A. africanus*, which is the ancestor of *Homo*, or that *A. afarensis* is not sufficiently different from *A. africanus* to be considered a distinct species.

These various views were recently forced into revision by the discovery by Alan Walker at Kenya's Lake Turkana of a 2.5 million-year-old, hyper-robust skull with a dish-faced profile and a cranial capacity of 410 cubic centimeters. It has been called the Black Skull because of its dark color, but is referred to by its museum number, KNM-WT 17000 (Kenya National Museum-West Turkana), in technical publications. Walker, Leakey, and colleagues think WT 17000 (not shown) is an early example of *A. boisei*, but Johanson, White, and colleagues want to assign it to *A. aethiopicus*, which they feel is the ancestor of *A. boisei*. *A. aethiopicus* is based on a lower jaw described from Ethiopia in 1967 by Camille Arambourg and Yves Coppens. It has received little recognition until now because it was an isolated mandible and no one really knew what to make of it. The lower jaw came from geologically older layers than WT 17000, and its locality was 90 miles north of the Kenyan specimen.

Johanson thinks that WT 17000 shares a few of the specialized features of *A. boisei*, such as large jaws and teeth and a huge bony ridge on top of the skull (sagittal crest) which provides attachment surfaces for the powerful chewing muscles needed to work the big jaws and teeth. However, because it is 500,000 years older than known specimens of *A. boisei*, WT 17000 should not be assigned to it. He further points out that WT 17000 shares many primitive characteristics with *A. afarensis* and should therefore be considered intermediate between *A. afarensis* and *A. boisei*. This scheme is depicted in Figure 36. It shows *A. afarensis* as the oldest common ancestor from which three lineages are derived. In eastern Africa we have *A. aethiopicus* and its descendant, *A. boisei*. In southern Africa there is *A. africanus*, giving rise to *A. robustus*. And finally, the *Homo* lineage emerges out of *A. afarensis*.

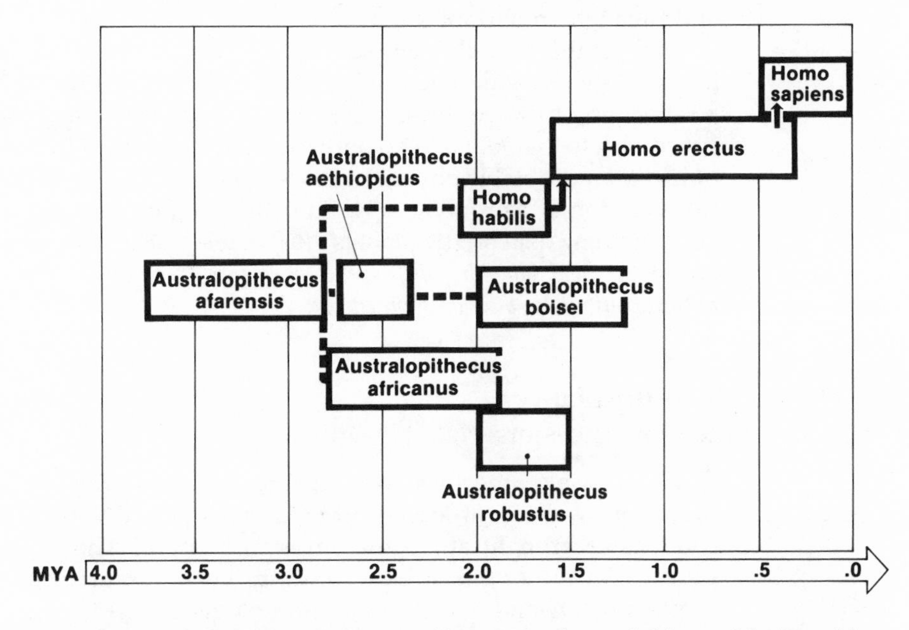

Figure 36. Tentative phylogeny of hominids, the model favored by Donald Johanson. Richard Leakey disagrees, maintaining that *A. afarensis* is not ancestral to the genus *Homo*, but that both *Australopithecus* and *Homo* have a common ancestor older than 6 million years that has yet to be discovered.

This scenario requires the least rearrangement from the views held prior to the discovery of WT 17000, but it is by no means the only interpretation. In fact, Kimbel, White, and Johanson (1988), in their review of the implications of WT 17000, cannot agree among themselves on a phylogeny, and they present four alternatives.

The dust has not yet settled from the WT 17000 discovery. Other views on australopithecine relationships can be found in Walker et al. (1986), Shipman (1986), and Delson (1987). So Figure 36 should be taken as tentative, and it will surely be revised in light of new fossil discoveries and new interpretations of known fossils. Perhaps an early *Homo* from Ethiopia or Kenya will eventually be found to fill the 700,000-year gap between *A. afarensis* and *H. habilis*. The search for fossils goes on, each new discovery bolstering or shaking the various hypotheses. The reconstruction of our ancestry is an exciting, ongoing enterprise, and the most important discoveries may be yet to come.

The fact that evolutionary scientists continuously revise their interpretations as more fossils and new data become available is a reflection of the youthful vigor of this field. This may be disconcerting to the public, who perhaps think in terms of the much older and more stable physical sciences, but is it so different from the situation of astronomers when confronted by Voyager with amazing new data on the planets? *Reinterpretation of our origins in the light of new data, the very essence of science, is an option not open to the creationists, by their own choice.*

The Origin of Bipedality
What Made Our Ancestors Walk Upright?

The australopithecines were the first upright walkers, and they were walking at least 4.0 MYA. What selective pressures could have brought about this unusual posture? The complex problem of the origin of bipedality has been speculatively addressed by Owen Lovejoy, an anthropologist at Kent State University. He suggested that the need to carry offspring and provisions was an important selective force for the development of upright walking.

Human infants are helpless during their long developmental period and are not as capable of clinging to their mothers as are young apes. The importance of bipedalism is that it frees the female's arms to carry the infant, thus protecting it from accidental injury or predation. That the males could now carry food to the females and offspring also contributed to the survival of the young.

Females, carrying their young in one arm, could forage for food with the other arm near the camp site, while the males hunted farther afield. This would save the female and child the rigors and dangers of prolonged travel and permit the female to build a deep familiarity with the resources of her home range. Competition between females and males for local food supplies would also be reduced. Food-sharing and division of labor could have resulted in long-term **pair bonding** between the male and female, which would ensure that the male would provision the female and their genetic offspring on a long-term basis. By ensuring that the infant is better fed and protected, the evolutionary fitness of both parents is enhanced, because their genes have a better chance to be passed along when the offspring reproduces.

Pair bonding carries with it the concept that the female no longer need show a recognizable fertile period. But that necessitates frequent copulation with her mate to ensure fertilization, which in turn strengthens the pair bond. The evolution of secondary sexual characteristics, such as the prominent pubic hair patterns, larger penis, and prominent female breasts further enhances the pair bond. The lack of an externally visible heat period in the female diminishes the attraction of other males and reduces cheating and male-male conflict over females. Polygynous apes such as gorillas and chimpanzees have large canine teeth, which the males use in fighting for females. Hominid canines are smaller than those of the apes but still slightly dimorphic. These tendencies toward monogamy, canine reduction, food-sharing, and bipedalism all seem to be part of the same evolutionary package.

There are many critics of Lovejoy's explanation. They cite the fact that parental care is not limited to monogamous primates, and the fact that the early hominids exhibited a marked sexual dimorphism, as expected in a polygynous species. Most likely, multiple factors contributed to the origin of bipedality.

The Genus *Homo* and Modern Humans

How Long Have There Been People?

There are three currently recognized species of *Homo*: *Homo habilis*, from 2.1 to 1.5 MYA; *Homo erectus*, from 1.6 to 0.25 MYA; and our own species, *Homo sapiens*, from 0.5 MYA to the present. These three species, overlapping in time, probably represent a sequence.

Homo habilis (2.1 to 1.5 MYA, Figure 37) was described by Louis Leakey (husband of Mary, father of Richard) and colleagues from Olduvai Gorge in eastern Africa, near the site where Mary Leakey had previously discovered *A. boisei*. *H. habilis* (the name means "handy man") is now known from southern and eastern Africa as well, and has been found in association with primitive stone tools. Because of its small cranial capacity, 657 cubic centimeters (cc), some anthropologists did not care to admit this species to membership in our genus, *Homo*. They felt it was an australopithecine. Richard Leakey's find at Lake Turkana, Kenya, in 1972 of skull 1470, virtually complete and with a cranial capacity of 775 cc, provided the brainpower necessary for *H. habilis* to be accepted in *Homo*. A few anthropologists, however, consider that *H. habilis* would more properly be designated *Australopithecus habilis*. This demonstrates, not confusion, but the transitional nature of the fossil record. Johanson and colleagues (1987) have reported on a 1.8-million-year-old *Homo habilis* from Olduvai. This find is notable because it demonstrates that this species, the first of the known *Homo* line, was quite small (3½ feet in the female) with very long arms. *H. habilis* was apparently rather similar to *A. afarensis*, which raises the interesting possibility that there was a relatively rapid increase in size, because just 200,000 years later *Homo erectus* stood 5 to 6 feet tall, and had proportionally shorter arms.

Homo erectus (1.6 to 0.25 MYA) has also been found at Olduvai, and seems to be the descendant of *H. habilis*. In eastern Africa,

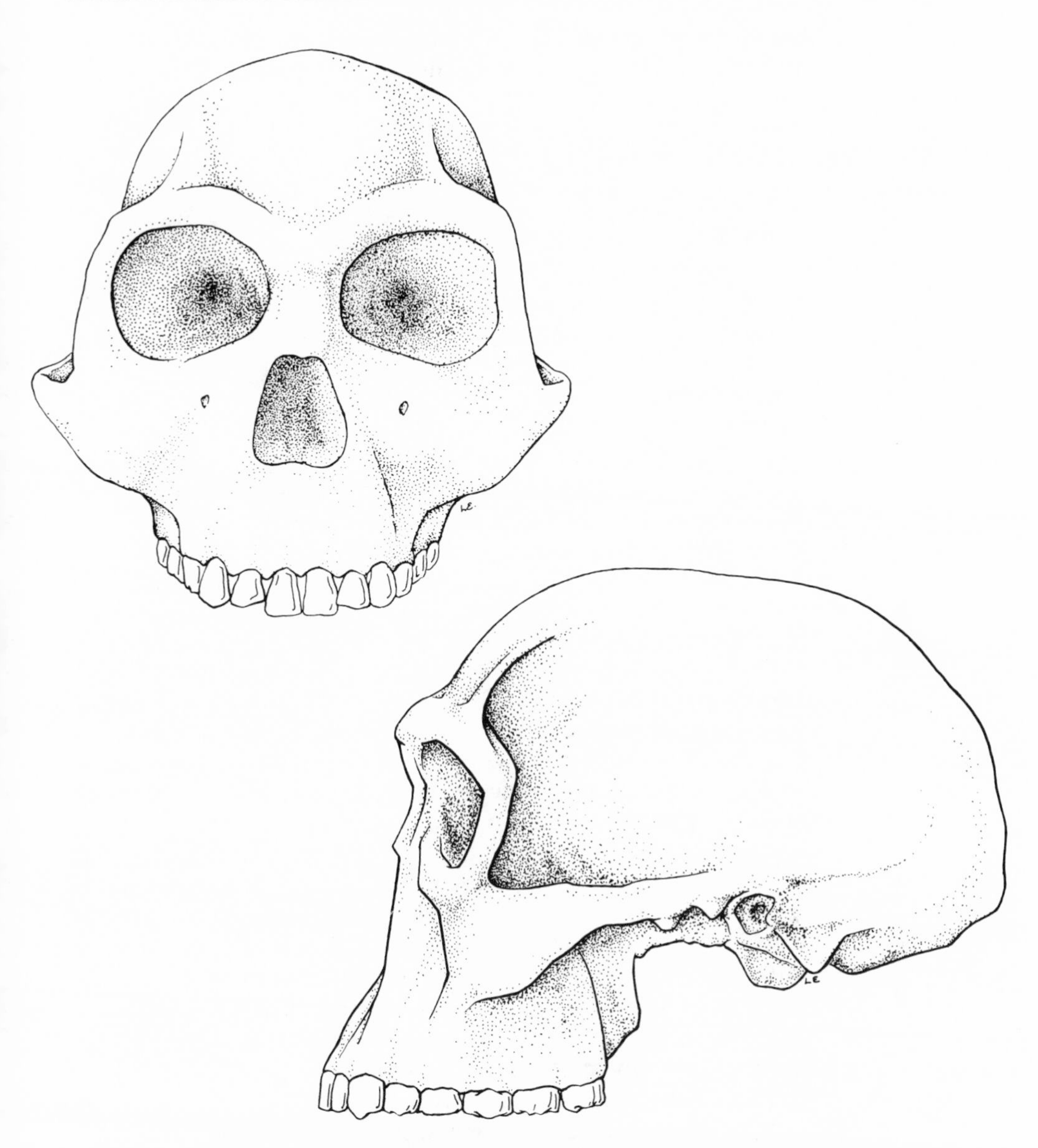

Figure 37. Partial reconstruction of *Homo habilis*, 1.8 million years old, from Koobi Fora, East Turkana, Kenya. Approximately one-half actual size.

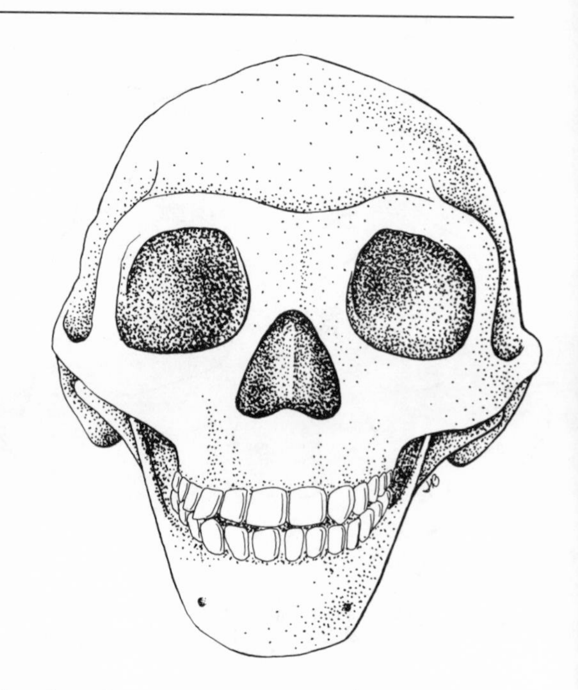

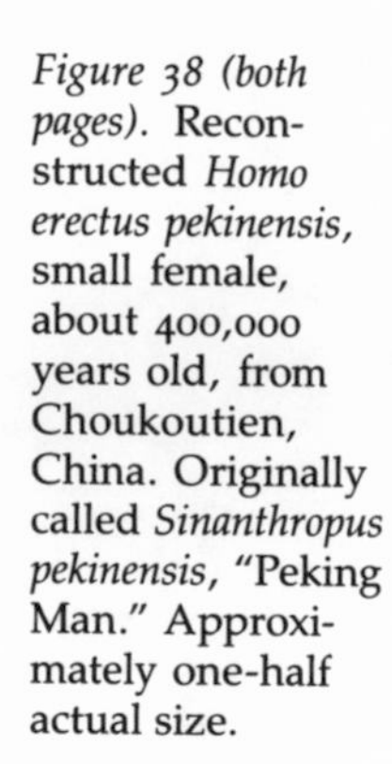

Figure 38 (both pages). Reconstructed *Homo erectus pekinensis,* small female, about 400,000 years old, from Choukoutien, China. Originally called *Sinanthropus pekinensis,* "Peking Man." Approximately one-half actual size.

early *H. erectus* coexisted with *A. boisei* and *H. habilis*. *H. erectus* (Java and Peking man [Figure 38] and others) had a cranial capacity of around 1,000 cc. The fossils of *H. erectus* display robust features with large jaws, well-developed brow ridges, a low, sloping forehead, and little or no chin. Their teeth were much smaller, reflecting a dietary shift from plants to meat, and their brain was larger than the *H. habilis* brain.

As far as we can tell from the fossils at hand, *Homo erectus* was the first hominid to leave the African continent and was widely distributed in Africa, Europe, and Asia. This species made stone and bone tools, employed fire, and lived and hunted cooperatively from a home base. Their brain was sufficiently complex to permit speech, but we will probably never be certain whether or not they actually spoke. *H. erectus* persisted until about 250,000 years ago in China and Java (Table 2, p. 102).

It is difficult to determine exactly when *Homo erectus* gave rise to our species, *Homo sapiens*. Some anthropologists put the transition as early as 500,000 years ago. Specimens called ar-

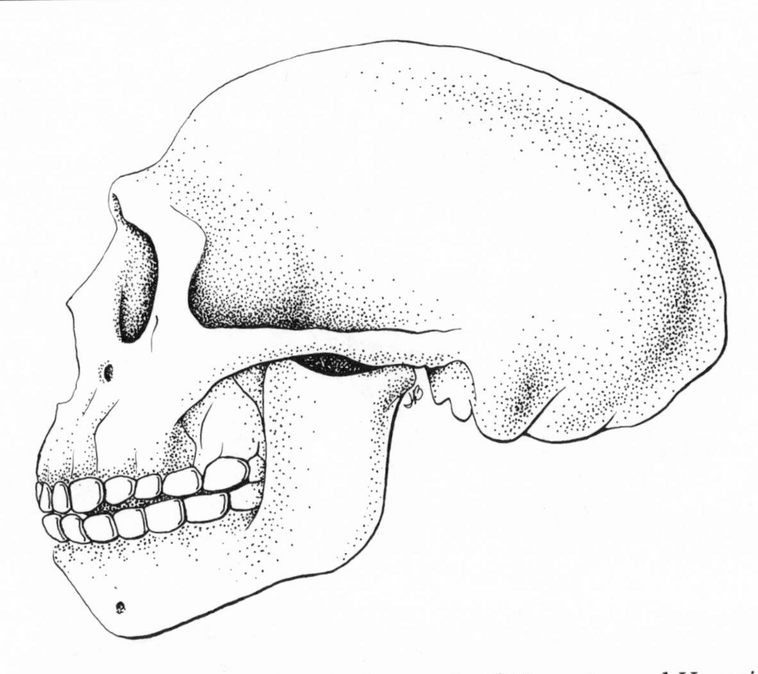

chaic *Homo sapiens*, showing a mosaic of *H. erectus* and *H. sapiens* characteristics, have been found at Mauer, West Germany (mandible), Petralona, Greece (skull), and Arago, France (face), and may be as old as 400,000 years. Other anthropologists feel that "sapientization" occurred about 250,000 years ago. The cranial capacity of archaic *Homo sapiens* (Table 2) ranged from 1,200 cc for specimens from Steinheim, West Germany, to 1,300 cc from Swanscombe, England. Compared with *H. erectus*, these specimens had heavy but smaller brow ridges, thinner skulls, and decreased tooth size. They were seemingly intermediate between *H. erectus* and the Neanderthal people.

The Neanderthals (named after the Neander Valley in Germany) emerged about 150,000 years ago and persisted until about 32,000 years ago. *Homo sapiens neanderthalensis* (Figure 39) is a member of our own species, but has been portrayed in a poor light in the older literature as a beetle-browed, shambling subhuman. Specimens from La Chapelle-aux Saints were deformed from arthritis and were incorrectly reconstructed in early resto-

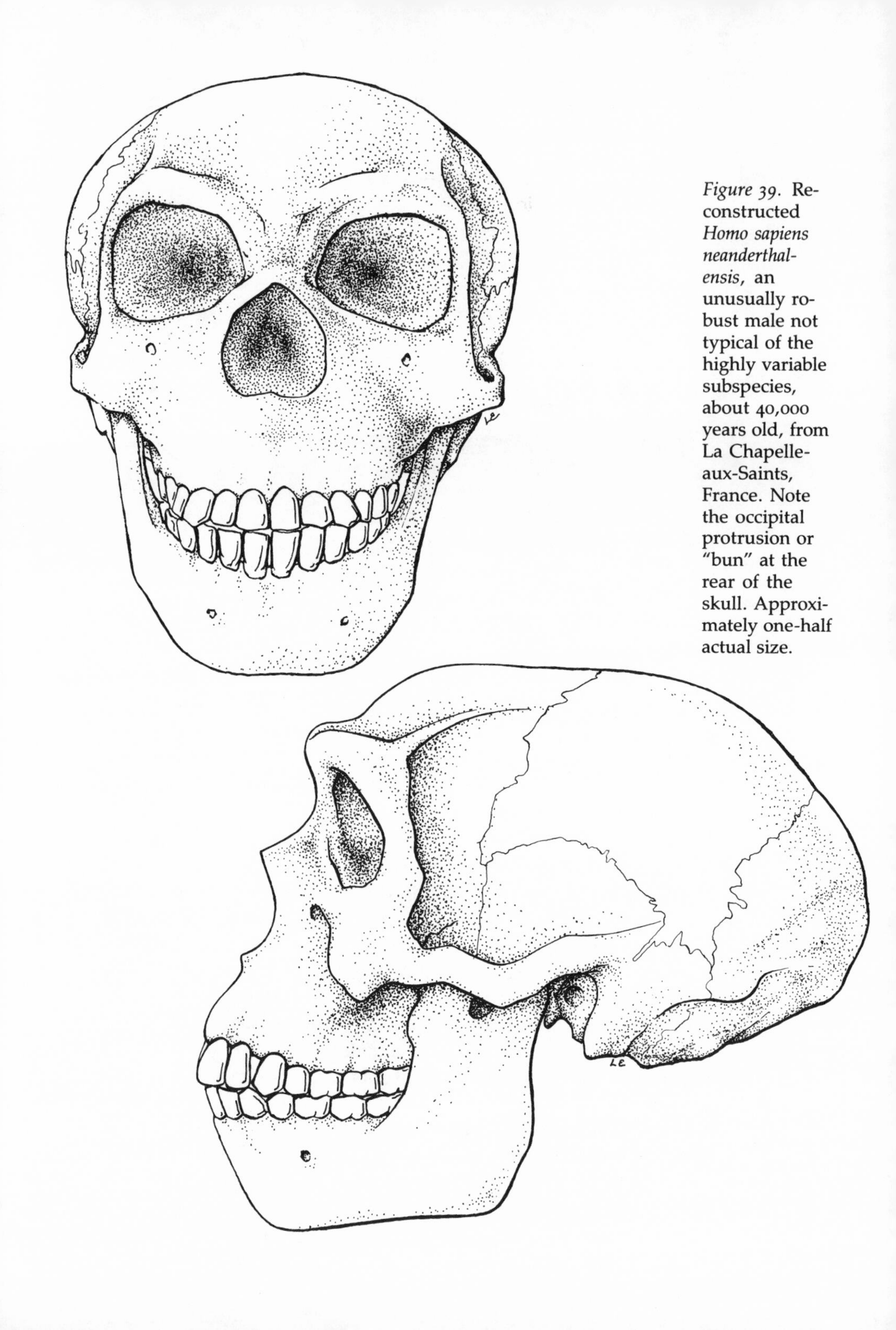

Figure 39. Reconstructed *Homo sapiens neanderthalensis*, an unusually robust male not typical of the highly variable subspecies, about 40,000 years old, from La Chapelle-aux-Saints, France. Note the occipital protrusion or "bun" at the rear of the skull. Approximately one-half actual size.

rations. Such errors contributed to the brutish image that is conjured up in the public's mind by the name Neanderthal. In fact, the Neanderthals were extremely varied and widely dispersed. They had an elaborate culture and may have been the immediate ancestors of modern *Homo sapiens*. They were more muscular, more barrel-chested, and shorter than modern humans, but if cleaned up, shaved, and dressed in business suits, they could probably pass for television evangelists. (For the rehabilitation of the Neanderthal image, see the cover stories of the October issue of *Science 81* and the May 1989 issue of *Discover*.)

The "classic" Neanderthal, which ranged widely in Europe and North Africa, had a large skull with heavy brow ridges, a weak chin, and prognathous (protruding) jaws. "Progressive" Neanderthals from the Middle East showed less massive features and more rounded skulls. Specimens from Mount Carmel in Israel and Shanidar in Iraq were intermediate between "classic" Neanderthals and modern humans.

Interpretation of the relationship between Neanderthals and modern humans may have to be revised in light of a recent paper by French scientist Hélène Valladas and colleagues. They reported on a primitive form of anatomically modern human fossils from the Qafzeh cave in lower Galilee. Charred flint flakes from the cave yielded a date of 92,000 years ago. This date is 50,000 years earlier than previous estimates for the appearance of modern humans. If correct, it means that some modern humans occupied southwest Asia before Neanderthals arrived. The implications for Neanderthal taxonomy are not yet clear, but Valladas et al. claim that their discovery excludes a close phylogenetic relationship between the Cro-Magnons and Neanderthals. However, other anthropologists see this new information as evidence that Neanderthals interbred with early modern humans and produced fully modern *Homo sapiens* about 40,000 years ago. For more details, see Valladas et al. (1988) and Stringer's (1988) comments, along with the summary in *Science News* (1988, vol. 133: 138).

Whatever their relationship with the Neanderthals, modern humans (Figure 40) emerged out of this extremely variable gene pool perhaps 40,000 years ago. Whether Neanderthals evolved

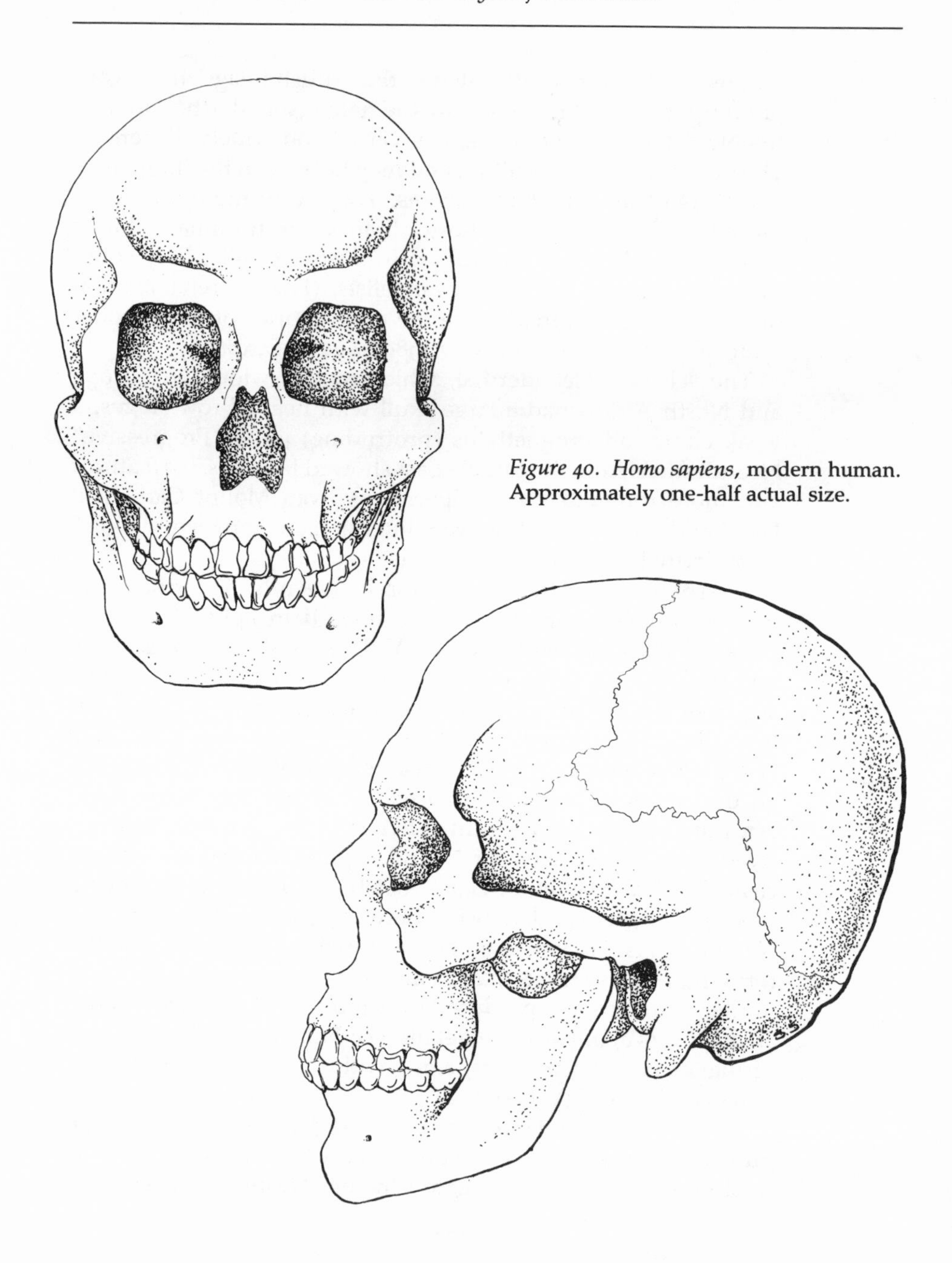

Figure 40. *Homo sapiens,* modern human. Approximately one-half actual size.

into modern humans, or whether modern humans displaced Neanderthals or interbred and genetically swamped them, is not clear, but by about 32,000 years ago fully modern human fossils had replaced the Neanderthals everywhere. In the strict archaeological sense, the term "Cro-Magnon" (pronounced "crow-mannion") applies only to the Upper Paleolithic people from southwestern France from about 35,000 to 10,000 years ago, but the term has also been broadly used to denote the first modern people regardless of locality. Their remains have been found on all continents. Cro-Magnon peoples reached these divergent localities at different times. They were the first known fully modern humans, and were a heterogeneous assemblage sharing considerable geographic variation, which is persuasively explained as adaptation to local environments. Their brow ridges and teeth were small, the chin prominent, and the forehead high; the rounded brain case averaged near 1,400 cc. The Cro-Magnons decorated their caves with paintings and carvings. *There is no denying that they are our immediate ancestors*.

The accelerating pace of hominid fossil discoveries is truly dazzling. In Darwin's time, only a few Neanderthal remains were known, and they were misunderstood. Today we have a whole cast of characters in the drama of human evolution. *These fossils are the hard evidence of human evolution*. They are not figments of scientific imagination. If the australopithecines, *Homo habilis* and *H. erectus*, were still alive today, and if we could parade them before the world, there could be no doubt of our relatedness to them. It would be like attending an auto show. If you look at a 1953 Corvette and compare it to the latest model, only the most general resemblances are evident, but if you compare a 1953 and a 1954 Corvette, side by side, then a 1954 and a 1955 model, and so on, the descent with modification is overwhelmingly obvious (Figure 41). This is what paleoanthropologists do with fossils, *and the evidence is so solid and comprehensive that it cannot be denied by reasonable people*. There are quibbles about individual relationships, but each new discovery helps fine-tune our increasingly detailed knowledge of human evolution.

Figure 41. Evolution of the Corvette. Everything evolves in the sense of "descent with modification," whether it be government policy, religion, sports cars, or organisms. The revolutionary fiberglass Corvette evolved from more mundane automotive ancestors in 1953 (above left). Other high points in the Corvette's evolutionary refinement include the 1962 model (below left), in which the original 102-inch wheelbase was shortened to 98 inches and the new closed-coupe Stingray model was introduced; the 1968 model (above right), the forerunner of today's Corvette morphology, which emerged with removable roof panels; and the 1978 silver anniversary model (below right), with fastback styling. Today's version (not shown) continues the stepwise refinements that have been accumulating since 1953. The point is that the

Corvette evolved through a selection process acting on variations that resulted in a series of transitional forms and an endpoint rather distinct from the starting point. A similar process shapes the evolution of organisms. (Photos courtesy of General Motors Corporation.)

5 Science, Religion, Politics, Law, and Education

The evolution/creation debate is ostensibly about science, but that is simply what the fundamentalists want us to believe, for a great deal more is at stake, and they are well aware that creationism is not science. The goals of "creation science" have far more to do with religion, politics, law, and education, and to whatever extent the scientific trappings of "creation science" are accorded credence, by the schools or by the courts, much will suffer: freedom of thought, an informed, open-minded American public, the vitality of science and technology, and the fate of our society in an increasingly competitive, increasingly educated world.

The Grim Realities of Science Education

How Bad, Really, Are Our Schools?

That Americans are in general poorly educated is not news, and our younger citizens, taken as a group, may have learned less than their parents or grandparents did. Our politicians routinely preach a commitment to education, but a look at some surveys demonstrates how hollow that commitment is. The National Geographic Society, for example, in the summer of 1988, released a report on (to quote *Newsweek,* August 8) "the dismal state of Americans' knowledge about the globe." Young Americans, "like their predecessors, still lust after adventurous trips to exotic places but now don't have a clue how to find them on a map." The survey compared U.S. geographic literacy with that of our neighbors Mexico and Canada and other industrialized countries (Sweden, West Germany, Japan, Italy, France, and the United Kingdom), on the basis of Gallup interviews "with representative samples of adults 18 and older—10,820 in all." Among the Americans, 14 percent could not even pick out the United States on a map, "and among 18- to 24-year-olds, Americans finished dead last" on the complete battery of questions.

This is a terrible indictment of American society, politics, and education, and we fare little better in our knowledge of science, biological or otherwise. A recent study conducted by the International Association for the Evaluation of Educational Achievement (*Science Achievement in Seventeen Countries: A Preliminary Report*, Pergamon, New York, 1988) compared students in grades 4/5 (ten-year-olds), 8/9 (fourteen-year-olds), and 12/13 (seventeen-year-olds), among various countries. American seventeen-year-olds ranked in the bottom 25 percent in biology, chemistry, and physics, behind students from England, Hong Kong, Singapore, Hungary, Poland, and Japan, to name a few. Out of seventeen countries, American fourteen-year-olds ranked fourteenth in science and mathematics, and American ten-year-olds finished in the middle of the group. Results from different American schools varied widely. The American showing was in fact similar to that of developing countries that display sharp contrasts between elite schools and other schools. The authors concluded the U.S. summary with the understatement, "For a technologically advanced country, it would appear that a re-examination of how science is represented and studied is required."

In a survey of undergraduates in three states (excerpted in the *Chronicle of Higher Education*, November 19, 1986, p. 37), more than half said they were creationists and one-third or more believed in ghosts, communication with the dead, extraterrestrials, aliens, Big Foot, etc. The prevalence of such pseudoscientific beliefs is an indictment of current science-education practices, which seem less effective than education via supermarket tabloids. The survey also revealed that those who accepted creationism were less likely to read books and had lower grade-point averages than the noncreationists. What that may mean is not altogether clear, but at the least it suggests an anti-intellectual suspicion of knowledge on the part of those sympathetic to creationism. See Harrold and Eve (1986) and Eve and Harrold (1986) for details of the study and Harrold and Eve (1987) for additional survey data and for an attempt to understand pseudoscientific beliefs about the past.

Several other opinion polls have revealed a public sympathy toward teaching creationism along with evolution at both the high school and the university level. Even university students fall prey to the equal-time argument, notwithstanding that the creationist interpretation of life is on a par with the idea of a flat Earth. In spite of a trend toward greater acceptance of evolutionary theory with increasing biological education, well over half the biology graduate students surveyed at The Ohio State University favored teaching creationism in public schools.

Another survey showed that only 12 percent of Ohio's high school biology teachers could select from five choices the phrase that best described the modern theory of evolution! This is pathetic testimony to both science education and teacher training. References to these surveys are given at the end of the text under Fuerst (1984) and Zimmerman (1986, 1987). In an equally alarming survey of 730 Ohio school board presidents (Zimmerman 1988), which brought 336 responses, 53 percent felt that "creation science" should be favorably taught in public schools, and most of this group felt it should be presented in biology or science classes. Less than 2 percent of the school board presidents were able to correctly select the statement that best described the theory of evolution from a list of five choices. Nearly 50 percent indicated they would do nothing if they learned that "creation science" was favorably taught in science class in their district, and 57 percent indicated that the school boards themselves should determine whether "creation science" or evolution should be included in science classes. The crisis in science education is certainly alive and well in Ohio.

So what became of the Columbus public schools' Biology Curriculum Guide that I mentioned in the Preface? As a result of protests by scientists and concerned citizens, it was withdrawn, and in October 1984 the Columbus School Board removed the teaching of creationism from science classes. But when the Board's new outline appeared in March 1985, the word "evolution" was also removed from the list of topics to be covered in biology class. We were assured that evolution would continue to be taught in the genetics unit of biology classes, but that it was prudent to avoid the "E" word. That the central unifying theory of a major discipline, accepted almost without exception in the rest of the educated world, must be smuggled in the back door of the classroom, for fear of offending the creationists and the fundamentalists, is an incredible commentary on the state of public education in the United States.

Following a speech to a fundamentalist coalition in Dallas in 1980, then Republican presidential candidate Ronald Reagan held a press conference at which he was asked if he thought the theory of evolution should be taught in public schools. He replied, "Well, it's a theory, it is a scientific theory only, and it has in recent years been challenged in the world of science and is not yet believed in the scientific community to be as infallible as it once was believed. But if it was going to be taught in the schools, then I think that also the biblical theory of creation, which is not a theory but the biblical story of creation, should also be taught" (*Science*, 1980, 209: 1214). One must wonder where the President got his scientific advice. Here is ignorance (and pragmatic politics) celebrated at the highest level through an anti-intellectual appeal to a voting constituency.

In just such ways, a great deal of misinformation has been propagated in the public media about the evolution/creation controversy. For this the scientific community must accept some blame. We have done a very poor job of explaining our work to the public, to the press, and even to biology students, as the dismaying recent surveys have shown. Scientists are usually deeply involved in research, and do not feel moved to spend time popularizing their work. Some scientists cherish the isolation of their ivory towers, and it is in any case difficult and time-consuming to explain elaborate technical theories in a way easily understood by the public. By failing to explain our research, however, we invite its misrepresentation at the hands of unqualified spokespersons.

Evolution vs. Religion
What Do the Creationists Hope to Achieve?

Science and religion are concerned with totally different spheres of human activity, and for the great majority of people—believers and nonbelievers alike—there is no basic conflict between them. In an encyclical in 1950 entitled *Humani Generis*, Pope Pius XII stated the Catholic position on evolution: a Catholic is free to accept any scientific theory about human origins provided it is acknowledged that, at some stage, God infused an immortal soul into the human body. This requires an act of faith, but not a denial of science, because science has nothing to say about gods and souls. Pope John Paul II, speaking to an audience of scientists and theologians in April 1985, echoed this position and urged continued scientific study, which he termed "serious and urgent."

Many evolutionary biologists are Catholic priests or nuns. Others are members of many of the world's other religions. They see evolution as God's plan, not as a denial of their belief in God, a view called **theistic evolution**. Most of the mainstream Protestant denominations such as Episcopalians, Presbyterians, and Methodists have also accommodated evolutionary theory. That they have done so without compromise to basic beliefs or principles is reflected in the fact that the biology departments of the universities operated by these religions teach the same evolutionary theory as the major state universities. In liberal theological circles it has become a cliché to state that, "The Bible was written to show us how to go to heaven, not how the heavens go."

This rather comforting view of theistic evolution has been disparaged by a Cornell University professor of the history of biology, William B. Provine, who claims that theistic evolution is no different than the Deism of the seventeenth and eighteenth centuries. But this is the playground of the philosophers, and I shall leave it to them. Provine has further elaborated his position in a book review published in *Academe* (1987, vol. 73: 50–52).

The only Christian religious groups that have problems with evolution are the Protestant fundamentalists, who insist on a literal interpretation of the Bible, much as Moslem fundamentalists insist on a literal interpretation of the Koran. Like Provine, Protestant fundamentalists reject theistic evolution, but for different reasons than Provine. It is the fundamentalists who are trying to infuse their brand of religion, taught as science, into public school classrooms. Having failed in the past to have evolutionary theory banned from the classrooms, the fundamentalists have adopted a new strategy, the promotion of a patchwork enterprise they call "creation science." Their goal is to have the biblical version of creation taught alongside evolution; and to disguise the essentially religious nature of creationism, they have dressed it up in scientific terminology. Since they have no persuasive arguments of their own, or even intuitively satisfying suggestions, their plan is to attack selected particulars of science and pretend to a science of their own.

A principal advocate of "scientific creationism" is a group of fundamentalists called the Creation Research Society (CRS). To be a member of the CRS one must have an advanced degree in some field of science and sign a statement of faith. The statement begins as follows: "1. The Bible is the written Word of God, and we believe it to be inspired throughout, all of its assertions are historically and scientifically true in all the original autographs. To the students of nature, this means that the account of origins in Genesis is a factual presentation of simple historical truths." The statement concludes with other points, involving God's direct creation of the Earth and all things in six days, Noah's flood, Adam and Eve, sin, and salvation through Christ.

The membership of the CRS consists mostly of engineers, chemists, aerospace workers, technicians, computer specialists, and such. Few legitimate biologists, geologists, or anthropologists are willing to sign such a statement of faith; and a degree in engineering, chemistry, or computer science scarcely qualifies a CRS member to speak with knowledge and authority about biology, geology, astronomy, or anthropology. The aim of the CRS is nonetheless to force the scientific evidence into compliance with a literal interpretation of the Bible. The arguments of these fundamentalist missionaries often involve tortured logic, a stubborn denial of the evidence, a shallow understanding, or a reckless disregard for the truth. Some of their favorite arguments and a brief scientific refutation of each are set out below. It can be seen that the creationists' arguments are not only antibiology but also anti-physics, anti-astronomy, and anti-geology. *In short, they reject all scientific knowledge that does not fit their view of the world.* They do not question the methods or philosophy that yield, say, the science of flight, for who could doubt that airplanes fly? But when the same methods and philosophy are put to the study of life and human origins—a subject the Bible does address—they question the integrity of science. The creationists fight a desperate, rear-guard action, seeking to increase their numbers while refusing to accept the obvious.

Some Creationist Claims
Do They Raise Any Legitimate Doubts?

1. *Evolution violates the second law of thermodynamics. Entropy (disorder) is always increasing. Since order does not arise out of chaos, evolution is therefore false.*

These statements conveniently ignore the fact that you *can* get order out of disorder if you add energy. For example, an unassembled bicycle that arrives at your house in a shipping carton is in a state of disorder. You supply the energy of your muscles (which you get from food that came ultimately from sunlight) to assemble the bike. You have got order from disorder by supplying energy. The Sun is the source of energy input to the Earth's living systems and allows them to evolve. The engineers in the CRS know this, but they permit this specious reasoning to be published in their pamphlets. Just as the more structured oak tree is derived from the less complex acorn by the addition of energy captured by the growing tree from the Sun, so sunlight, via photosynthesis, provides the energy input that propels evolution. In the sense that the Sun is losing more energy than the Earth is gaining, entropy is increasing. After death, decay sets in, and energy utilization is no longer possible. That is when entropy gets you. What does represent an increase in entropy, as biologists have pointed out, is the diversity of species produced by evolution.

2. *The small amount of helium in the atmosphere proves that the Earth is young. If the Earth were as old as geologists say, there would be much more helium, because it is a product of uranium decay.*

Helium, used to suspend blimps in air, is a very light gas and simply escapes into space; like hydrogen, it cannot accumulate in Earth's atmosphere to any great extent.

3. *The rate of decay of the Earth's magnetism leads to the calculation that the Earth was created about 10,000 years ago.*

The Earth's magnetic field does indeed decay, but it does so cyclically, every few thousand years, and it is constantly being renewed by the motion of the liquid core of the Earth. The "fossil magnetism" recorded in ancient rocks clearly demonstrates that polar reversals (shifts in the direction of the Earth's magnetic field) have occurred both repeatedly and irregularly throughout Earth history; the calendar of these reversals was established over two decades ago, and quickly became the linchpin in the emerging theory of plate tectonics and continental drift.

4. *If evolution were true, there would have to be transitional fossils, but there are none; therefore, evolution did not occur.*

There are many transitional fossils, including the ape-human transitional form, *Australopithecus*. *Eusthenopteron* shows marvelous intermediate characteristics between the lobe-finned fishes and the amphibians. The transitional fossils between amphibians and reptiles are so various and so intermediate that it is difficult to define where one group ends and the other begins. *Archaeopteryx* is clearly intermediate between reptiles and birds. In spite of such reptilian affinities as a long bony tail, toothed jaws, and clawed wings, creationists declare that because *Archaeopteryx* had feathers, it was a bird, not a transitional stage between reptiles and birds. Having no explanations of their own, the creationists attempt to deny the transitional fossils out of existence.

5. *Fossils seem to appear out of nowhere at the base of the Cambrian; therefore, they had to have been created.*

The earliest microfossils date back, in fact, to the Precambrian, about 3.5 billion years ago. A variety of multicellular life appears in the fossil record about 670 million years ago, which is 80 million years before the Cambrian. The Cambrian does seem to explode with fossils, but that is simply because the first shelled organisms, such as the brachiopods and the trilobites, date from the Cambrian; their resistant shells fossilize far more readily than their soft-bodied ancestors of the Precambrian. What is more, Precambrian rocks are so old that they have been subjected to a great deal of deformation. We are thus fortunate to have *any* Precambrian fossils of soft-bodied animals. Still more fossils are discovered every year, and each one further weakens the creationist position.

6. *All fossils were deposited at the time of the Noachian flood.*

There is not a shred of evidence in the geological record to support the claim of a single, worldwide flood. Geological formations such as mountain ranges and the Grand Canyon require millions of years to form, and the fossil record extends over several billion years. The time required for continents to have drifted into their present positions is immense. These things cannot be accounted for by a single flood lasting a few days or years.

7. *There are places where advanced fossils lie beneath more primitive fossils.*

Earth movements such as faulting and thrusting produce these discontinuities; the older rock has simply been pushed over on top of the younger rocks, as we sometimes see even along highway cuts. These places are easily recognized and explained by geologists. They cannot be explained by the creationists' belief that all fossils are the result of the Noachian flood. Thus the creationists' attempt to fault evolutionary theory by these means ends up demolishing one of their own pet claims.

8. *The chances of the proper molecules randomly assembling into a living cell are impossibly small.*

Simulation experiments have repeatedly shown that amino

acids do not assemble randomly. Their molecular structure causes them to be self-ordering, which enhances the chances of forming long chains of molecules. Simulation experiments also demonstrate that the formation of prebiotic macromolecules is both easy and likely and does not require DNA, which is a later step in the evolution of proteins. The stepwise application of cumulative natural selection acting over long periods of time can make the improbable very likely.

9. *Dinosaur and human footprints have been found together in Cretaceous limestone at Glen Rose, Texas. Therefore, dinosaurs could not have preceded humans by millions of years.*

This Fred Flintstone version of prehistory is one of the most preposterous and devious claims that the fundamentalists make, and they have made it in both books and films. The "man-tracks" seen by creationists stem from two sources. One is wishful imagination, whereby water-worn scour marks and eroded dinosaur tracks are perceived as human footprints. The other source is deliberate fraud. Creationist hoaxers obscure the foot pads of dinosaur tracks with sand and photograph what remains, the dinosaur's toe impressions. When reversed, the tip of the dinosaur toe or claw becomes the heel of a "human" print. These prints are shown in poor-quality photographs in creationist literature and films. Because the stride length (7 feet) and foot length (3 feet) exceed any possible human scale, the fundamentalists call these the giants mentioned in Genesis. In addition to doctored dinosaur tracks, there are other hoaxed prints circulating in this area of Texas. In fact, carved footprints were offered for sale to tourists in curio shops during the Great Depression. These caught the eye of the paleontologist Roland T. Bird, who recognized them as fakes, but they eventually led him to the legitimate dinosaur footprints at Glen Rose. This area has since been extensively studied by paleontologists, and numerous species of reptiles and amphibians have been catalogued. No genuine human tracks exist there, but by leading to genuine new discoveries, the hoax became a boon to science.

10. *Biologists have never seen a species evolve.*

On a small scale, we certainly have. Using allopolyploidy and artificial selection, scientists have manufactured crop plants and horticultural novelties that are reproductively isolated from the parental stock. In addition, one can see stages of incipient speciation in nature by looking at clinal variations and subspecies, that is, gradual change in the characteristics of a population across its geographical range. Major evolutionary changes, however, usually involve time periods vastly greater than man's written record; we cannot watch such changes, but we can deduce them by inference from living and fossil organisms.

11. *Evolution, too, is a religion, and requires faith.*

Creationists are beginning to admit that their "science" is not science at all, and that it depends on faith, but, they are quick to add, so does evolution. Not so. Biologists do not have to *believe* that there are transitional fossils; we can examine them in hundreds of museums around the world, and we make new discoveries in the rocks all the time. Scientists do not have to *believe* that the solar system is 4.5 billion years old; we can test the age of Earth, Moon, and meteoritic rocks very accurately. We do not have to *believe* that protocells can be easily created from simple chemicals in the laboratory; we can repeat the experiments, with comparable results. We can also create artificial species of plants and animals by applying selection, and we can observe natural speciation in action. That is the big difference between science and religion. Science exists *because of* the evidence, whereas religion exists upon faith—and, in the case of religious fundamentalism and creationism, *in spite of* the evidence.

12. *The number of humans today would be much greater if we have been around as long as evolutionists say we have.*

This notion makes some very naive assumptions about birth and death rates, and the fecundity of early humans, and assumes that populations are always growing, when in fact most animal populations are at a level somewhat lower than the carrying capacity of their environment. Such stable populations remain stable for long periods of time, held in check by environmental constraints. It is only our own species' recently acquired ability to modify our environment that has allowed our numbers to get dangerously out of control. Ironically, it is our ability to master the environment—as the Bible commands us to do—that may yet do us in.

13. *The current rate of shrinkage of the Sun proves that the Earth could not be as old as geologists say, because the surface of the Sun would have been near the Earth's orbit just a few million years ago.*

This simplistic view neglects the fact that stars, such as our Sun, have life cycles during which events occur at different rates. The characteristics of a newly formed star are quite different from those of stars near death. Astronomers can see these differences today by observing young, middle-aged, and old stars. By now, we know a great deal about the Sun, and we know that it has not been shrinking at a constant rate.

14. *A living freshwater mussel was determined by Carbon 14 dating to be over 2,000 years old; therefore Carbon 14 dating is worthless.*

When used properly, Carbon 14 is a very accurate time-measuring technique. The mussel in this example is an inappropriate case for ^{14}C dating because the animal had acquired much of its carbon from the limestone of the surrounding water and sediment. These sources are very low in ^{14}C, owing to their age and lack of mixing with fresh carbon from the atmosphere. Therefore a newly killed mussel in these circumstances has less ^{14}C than, say, a newly cut tree branch. The reduced level of ^{14}C yields an artificially older date. The ^{14}C technique has no such problems with the tree branch that gets its carbon from the air, or with the campfire sites of ancient peoples. As with arcwelding or Cajun cooking, one must understand the technique to use it properly. This is another example of the self-correcting nature of science.

15. *The influx of meteoritic dust from space to Earth is about 14 million tons per year. If the Earth and Moon were 4.5 billion years old, then there should be a layer of dust 50 to 100 feet thick covering their surfaces.*

This estimate of dust influx is simply out of date. Space probes have found that the level of dust influx from space is about 400 times less than that. Creationists are aware of the modern measurements, but they continue to use the incorrect figure because it suits their purpose. Such is their honesty and scholarship.

Do these people believe that the astronauts would have been allowed to land on the Moon if NASA thought they would sink into 100 feet of dust?

16. *Prominent biologists have made statements disputing evolution.*

The out-of-context quote is one of the most insidious weapons in the creationists' arsenal, and reflects the desperation of their position. Biologists do not deny the *fact* of evolution. We do, however, debate its *mechanisms* and *tempo*. The debate reflects the vigorous growth of a major scientific concept; it is what goes on routinely in all healthy, growing branches of scholarship. Creationists dishonestly portray this as a weakness of the theory of evolution.

These 16 points are just a few of the creationists' arguments. There are others, but they are all of the same character—scientifically inaccurate, willful, or devious.

The Scopes Trial
Why Did This Modest Case Shake the Country?

The history of creationism in America has been reviewed by Numbers (1982). He traced its origins from early twentieth-century fundamentalist roots through the Scopes trial to the foundation of the Creation Research Society and its current incarnation as the Institute for Creation Research. The following account of the Scopes trial has been gleaned in part from Ruse (1982).

In the South, following the Civil War, several movements were formed to stem the influence of Bible-threatening science in schools. Following World War I, conservative Christians, spurred on by their success in establishing Prohibition, turned their attention to what was being taught in science classes. Several states passed anti-evolution laws, owing to pressure from groups called "fundamentalists" (because they adhered to Bible-affirming fundamental principles expounded at the 1895 Niagara Bible Conference).

Tennessee enacted legislation that prohibited the teaching of evolution in schools. A young biology teacher in the Dayton, Tennessee, High School, John T. Scopes, with support from the American Civil Liberties Union, allowed himself to be prosecuted under the anti-evolution law in 1925. Scopes was defended by the brilliant lawyer Clarence Darrow and prosecuted by three-time presidential candidate William Jennings Bryan, himself a fundamentalist and a great orator. The judge ruled that expert scientific witnesses for the defense would not be allowed, which left Darrow to fend for himself. The court took on a circus atmosphere when Darrow cross-examined Bryan on the literal truth of Genesis, and Bryan eventually conceded that the world was far more than 6,000 years old. The stress of the trial undoubtedly contributed to Bryan's death a few days after the trial.

Scopes was found guilty of teaching evolution in violation of Tennessee law (which, of course, he did do) and was assessed the minimum $100 fine. The verdict was overturned on appeal, due to a technicality, but in spite of the guilty verdict, the evolutionists were perceived as having won a major moral victory. Tennessee was held up for ridicule to the entire nation, owing in large part to the caustic newspaper reporting of H. L. Mencken. The affair caused other states to quietly shelve or dismiss proposed "monkey laws," and in 1968, in Epperson v. Arkansas, the Tennessee law and similar laws in Mississippi and Arkansas were declared unconstitutional by the U.S. Supreme Court.

An eyewitness account of the Scopes trial by a deeply religious and much respected geologist, Kirtley F. Mather, was republished by the Tennessee Academy of Science in 1982 and is cited in the references at the back of the book. The Scopes trial was recreated in the 1960 film *Inherit the Wind*, directed by Stanley Kramer and based on the play by Jerome Lawrence and Robert Lee. The movie stars Spencer Tracy as Darrow, Frederic March as Bryan, and Gene Kelly as Mencken. It is a powerful film classic worthy of a second look, and in fact a new production, starring Jason Robards as Darrow, Kirk Douglas as Bryan, and Darren McGavin as Mencken appeared on television in 1988.

After the Scopes trial, creationists channeled their energies inward, organized various creation research groups, and re-emerged in their current guise as "scientific creationists" demanding equal time with evolution, rather than the omission of evolution from the curriculum of public schools. This tactic led ultimately to the Arkansas trial.

The Arkansas Balanced Treatment Act
What Do the Creationists Say in Court?

The 1981 trial of Arkansas Act 590, the "Balanced Treatment Act," which required "creation science" to be taught in public schools along with evolution, provided an opportunity to see creationists' scholarship and their case at their best. (See Roger Lewin's account in *Science*, 1982, 215: 142–46, for more details; the following summary is based on his report.) A witness for the fundamentalists, Norman Geisler, of the Dallas Theological Seminary, testified that "it is possible to believe that God exists without necessarily believing in God." He also declared that UFO's were agents of Satan. Another fundamentalist witness, Henry Voss, a California computer expert, was withdrawn by the creationists at the last minute when he expounded upon demons at a pretrial deposition.

Many of the "creation scientists" admitted in pretrial depositions that what they practice is not scientific. Harold Coffin, of Loma Linda University (a Seventh-Day Adventist college), stated, "No, creation science is not testable scientifically." Ariel Roth, also of Loma Linda, when asked if "creation science" was really science, said, "If you want to define 'science' as testable, predictable, I would say no." (That a proposition be testable or predictable *is* the definition of science.) Remember, these are people who support the fundamentalist position; they were handpicked by the creationists to defend that position; but they cannot bring themselves, in a court of law, to claim that "creation science" is really science. Their honesty is refreshing.

Roth, not a member of the CRS, was presented as an expert on coral reefs whose thesis is that corals grow very rapidly and do not need millions of years to form massive reefs. He testified for 70 minutes, but the cross-examination was brief. Q: "What is the last sentence of your article on the growth of coral reefs?" A: ". . . this does not establish rapid growth of coral development." Q: "Is there any evidence that coral reefs were created in recent times?" A: "No." Q: "No further questions."

Coffin testified to the usual creationists' position of sudden appearances of complex organisms in the Cambrian, the absence of transitional fossils, etc. The cross-examination pointed to his scientific credibility. Q: "You have had only two articles in standard scientific journals since getting your Ph.D. in 1955, haven't you?" A: "That's correct." Q: "The Burgess shale (a geological formation in the Canadian Rockies with exceptionally well preserved marine fossils) is said to be 500 million years old, but you think it is only 5,000 years old, don't you?" A: "Yes." Q: "You say that because of information from the scriptures, don't you?" A: "Correct." Q: "If you didn't have the Bible, you could believe the age of the Earth to be many millions of years, couldn't you?" A: "Yes, without the Bible."

Five of the scientific witnesses called by the defense were members of the Creation Research Society (CRS), and thus had signed the statement affirming their belief in a literal interpretation of the Bible. Wayne Friar, a zoologist from King's College (a small Christian school in New York State), devoted most of his testimony to reading from outdated books on evolution published in 1929, 1930, and 1953, to demonstrate that the authors had misgivings about evolution. Upon cross-examination, Friar said he had signed the CRS statement of faith and a similar one as a condition of employment at King's College. He also admitted that a great deal has happened in the science of biology since 1929. Friar was willing to concede that a limited amount of evolutionary change is possible "within kinds." He could not adequately define "kinds," a term that appears in Genesis but not in scientific writing. Fundamentalists use the word "kinds" for every taxonomic category from species, genus, and family to order. He was asked, Q: "You believe the choice between evolution and creation is a matter of faith, don't you?" A: "Basically, yes."

Margaret Helder, a botanist and vice president of the CRS,

was shown to have published only one paper in non-creationist literature since 1971, and she stated in her deposition that there was no scientific evidence for special creation.

Donald Chittick, a physical chemist and member of the CRS, testified about geological evidence for Noah's flood, rapidly forming coal, and the invalidity of radiometric dating. He asserted the helium argument mentioned above. The cross-examination, which showed him to believe that the world was created in six natural days, hit directly at Chittick's lack of credentials. Q: "You have had no formal course in radiometric dating for 20 years, have you?" A: "Not since then." Q: "You have had only one article in a refereed journal since 1960, isn't that correct?" A: "Correct." (Lest the reader question this attention to numbers of published articles, it should be pointed out that working scientists, those who truly advance human knowledge, usually publish dozens of papers in comparable spans of years.)

There were two other witnesses for Act 590; both left the by-now familiar impression that the scientific evidence for creation is nonexistent. One of them, Chandra Wickramasinghe, a mathematician and an associate of Fred Hoyle in the *Archaeopteryx* fiasco discussed in Chapter 2, stated that no rational scientist would believe the Earth is less than one million years old, or that the world's geology could be explained by a worldwide flood.

The late Federal District Judge William Overton, of Little Rock, had no choice but to rule that Arkansas' Balanced Treatment Act was unconstitutional. His ruling stated that the act "was simply and purely an effort to introduce the biblical version of creation into the public school curricula." As such, it violated the First and Fourteenth Amendments of the U.S. Constitution.

The Louisiana Creationism Act
Why Do We Separate Church and State?

A similar law in Louisiana, the Creationism Act of 1981, was struck down on January 10, 1985, when Federal Judge Adrian Duplantier of New Orleans ruled that the law was unconstitutional on First Amendment grounds "because it promotes the beliefs of some theistic sects to the detriment of others." The law had required that creationism be taught when evolution is taught in public schools (because of legal challenges, the law had never been implemented).

On July 8, 1985, the 5th U.S. Circuit Court of Appeals upheld Duplantier's ruling, and in December the same court refused to reexamine the decision. This case, known as Edwards v. Aguillard, was then appealed to the U.S. Supreme Court. Seventy-two Nobel laureates and 24 scientific organizations filed a friend-of-the-court brief against the law.

On June 19, 1987, the Supreme Court, by a vote of 7 to 2, struck down Louisiana's Creationism Act. Justice William J. Brennan, writing for the court, said that the purpose of the Louisiana law "was to restructure the science curriculum to conform with a particular religious viewpoint." He added, "Because the primary purpose of the Creationism Act is to advance a particular religious belief, the act endorses religion in violation of the First Amendment." He further noted, "Forbidding the teaching of evolution when creation science is not also taught undermines the provision of a comprehensive scientific education."

The two recent Reagan appointees to the court, Chief Justice William H. Rehnquist and Justice Antonin Scalia, dissented. Scalia wrote the minority opinion and conceded that the law would be unconstitutional if there were truly nothing scientific to be taught under the rubric of "creation science." He felt, however, that "creation science" is a body of scientific knowledge.

Students and the public in general need to understand the difference between *ideas* based on religious beliefs and *scientific theories* based on evidence and reasoning. The appeal of creationism to a large portion of the population, and the fact that people such as Justice Scalia would give any validity at all to creationism, indicate how poorly the public understands science. The success of the creationists in building such extensive support is also a direct reflection of the failure of educators to make the facts known.

Put more emphatically, we educators and biologists cannot in good conscience do much gloating over the Supreme Court's decision; for except for our failure, how could such an unworthy challenge ever have attained the status and celebrity of a serious court case? Can one imagine physicists dancing in the streets because a court decided to allow them to teach the theory of relativity? Or geographers spiking the globe in the end zone after defeating the flat-earthers in court? If any of us has still not understood the gravity of the situation, we have only to realize that nonsense, in 1987, commanded enough influence to force a Supreme Court decision. And the decision was not even unanimous.

The Inevitable Triumph of Knowledge
Where Will It All End?

At least in a scientific context, legitimate scientists do not take the creationists or their claims seriously. Creationists have no data of their own to prove their assertions. In matters relevant to the origin and nature of life on Earth, they pursue no pertinent research worthy of the name. They studiously overlook the more telling modern scientific literature, except to extract out-of-context quotes. Their scholarly publication record is nonexistent. With all their talk about "scientific creationism," the creationists have produced no documentation in the scientific literature. No empirical, experimental, or theoretical evidence for "scientific creation" has been published in peer-reviewed science journals, the traditional method whereby scientists communicate the results of their research. Scott and Cole (1985) surveyed 68 likely journals for creationist manuscripts. Out of 135,000 submissions from 1980 to 1983, only 18 (0.01 percent) dealt with scientific support for creationism. Of these 18 articles, 12 were submitted to one *science education* journal (not the appropriate place for genuinely new findings). Fifteen of the manuscripts were rejected on grounds of poor scholarship. Three were still under review during the survey. (Dr. Scott has informed me that the remaining three manuscripts were subsequently rejected.) Obviously there is no such thing as "scientific creationism" in the professional literature of science. It exists only in the house publications of creationist organizations, and what appears in such places is empty of evidence, intellectual reasoning, or persuasive argument. Since the survey was of *submissions*, not of *acceptances*, creationists cannot claim that science journals are biased against them. It is clear that the creationists have nothing of scientific value to submit.

The entire case of the creationists thus consists of trying to disprove evolution. Their attacks *against evolution* are somehow supposed to *prove creationism*. Beginning philosophy students should immediately recognize this exercise in anti-logic.

In a political sense, however, one must take the creationists very seriously. They are concerned with achieving their goal (presenting fundamentalist beliefs in science classrooms), not with truth. They know the value of publicity, the unconscious appeal of the notion that where there is smoke (a public trial) there must be fire (a valid position)—for bad publicity is better than no publicity. Having failed to legislate their beliefs in the Arkansas and Louisiana cases, fundamentalists have turned their attention to local school boards and textbook selection committees in an effort to get their version of religion taught as science. Creationists will attempt to prevail by intimidating local board members, teachers, and textbook publishers into watering down science content. Parents who want to be sure their children are receiving a proper science education should take the trouble to find out what is being taught in their school system.

The proper place for the study of religious beliefs is in a church or temple, at home, or in a course on comparative religions, but not in a biology class. There is no place in our world for an ideology that seeks to close minds, force obedience, and return the world to a paradise that never was. It is unfair to the students, their teachers, and the nation as a whole for the students to be taught beliefs and misinformation instead of factual, testable theories in science class. Students should learn that the universe can be confronted and understood, that ideas and authority should be questioned, that an open mind is a good thing. *Education does not exist to confirm people's superstitions, and children do not learn to think when they are fed only dogma.*

There is no law that mandates the teaching of evolution, and there should not be, yet it is practically universally taught in universities and colleges around the world. The theory of evolution is what is taught because it is what best explains the data in a rational manner. The National Academy of Science, The American Association for the Advancement of Science, The National Association of Biology Teachers, and 72 Nobel Prize winners have all gone on record as supporting evolution and rejecting the teaching of creationism in science classes.

Biologists do not derive any special benefits from their espousing of evolution, and they do not set out to undermine anyone's faith in a supreme being. They are simply seeking to understand how the universe operates. Science is highly competitive; and because its theories are by definition open to challenge, science is also self-correcting. Any biologist who could disprove the theory of evolution would instantly become famous, possibly a Nobel laureate, and probably very wealthy. The fact that biology has been so spectacularly successful in describing and explaining the structure, processes, and diversity of the living world points to the validity of its theories. For example, such new techniques as tissue typing for organ transplants, the cloning of commercially valuable organisms, and genetic engineering whereby a gene for a human product such as insulin is inserted into bacteria are direct descendants of evolutionary theory. The increase in production from livestock and crops is based upon principles of selection. The kinship of humans and other animals is inferred from the fact that drugs are routinely tested in our mammalian relations before use on us. The rise of organisms resistant to drugs and pesticides is an example of evolution in action.

Evolution makes sense of the commonplace observation that organisms appear to be related, that more primitive groups show up in the fossil record before the more derived ones, that transitional fossils are difficult to classify, that mammalian embryos, including our own children, pass through a stage when they are equipped with the pharyngeal gill pouches of the lower vertebrates, that the limbs of all four-footed animals are built on the same plan, that vestigial organs exist in many animals including humans, that many essential biochemical pathways are common to all life, the essential unity of which is evident from the near-universal presence of DNA as the genetic coding molecule. The explanatory power of the theory of evolution led the eminent geneticist Theodosius Dobzhansky to remark, "Nothing in biology makes sense except in the light of evolution."

We do not know all there is to know about evolution. A great deal more remains to be learned, *but as more data are added, the theory itself evolves*. The theory of evolution does not need the approval of the fundamentalists, and no scientific theory is validated by ratification. Truth does not give way for legislation *or* for flights of fancy. The evidence is out there in the world of nature for all who are not blinded by religious preconceptions. Biologists, whose job it is to study life, have no doubt that evolution has occurred and is occurring, for evolution is a scientific theory that explains the facts.

But in another sense, evolution is itself a fact. Facts are individual data points—real things, real events in the real world. The facts of evolution described in this book exist regardless of the theoretical mechanisms we construct to explain them. The evidence is too vast and too varied to deny. *How* evolution occurs, via natural selection and the other processes mentioned in this book, is theoretical.

Creationists have deified their interpretation of a book, and they expect all to worship at their altar. They would have us substitute blind faith for reason. Because they insist on a literal interpretation of that book (most Christians and Jews do not), they are forced to discard all of modern biology, of which evolution is the cornerstone, plus most of geology and a good deal of physics and astronomy. Creationists insist on a young Earth, but if there is anything science knows, it is the great antiquity of the Earth and the solar system. This has been verified by at least five independent radioisotope clocks. In thousands of tests, rocks dated by three or more independent clocks yielded ages in good agreement. These dating procedures are based on the same theories as are nuclear reactors and bombs. There is no doubt that the reactors react and the bombs explode. The data from astrophysics establish distances and time scales consistent with the data from evolving stars. Light traveling at 186,000 miles *per second* takes 100,000 *years* simply to cross our own galaxy. The decomposition of organic matter to petroleum is known to require millions of years. The movement of continental plates, which results in the formation of oceans and mountains, requires millions of years. Life has been evolving for billions of years.

In light of the track record of scientific accomplishments, it is astonishing that the fundamentalists would assume that the world's biologists, geologists, physicists, and astronomers are all wrong. They must think a devious creator is playing games with us by providing sequential and transitional fossils in rocks that can be dated. They must think God is testing us by making it only appear that the older rocks are at the bottom and younger rocks are at the top of a stratigraphic column. If the Earth is only a few thousand years old, a devious god must have created the light that is already on its way to Earth from sources millions of light-years away. And as J. B. S. Haldane pointed out, God must have an inordinate fondness for beetles to have created over 250,000 different species.

Why would a creator make 2,000 species of fruit flies and put one-fourth of them only in the Hawaiian Islands? Were Adam and Eve created with all of man's parasites, including syphilis, herpes, and AIDS, already in place? Surely these things are much more rationally explained by evolution. Creationism has no explanatory powers, no application for future investigation, no way to advance knowledge, no way to lead to new discoveries. As far as science is concerned, creationism is a sterile concept.

The reason fundamentalists have such a deep hatred and fear of evolution is twofold. First, they mistakenly assume that science is inherently antireligious, that it is out to undermine their religious beliefs. Science is not attempting any such thing. It is saying that such religious beliefs cannot be considered as science, and that, when religious beliefs conflict with what science knows, it is not rational to assume that the science is wrong. Insisting on a literal interpretation of the Bible is a decision fundamentalists have made, not one that was thrust upon them, and a literal interpretation is one that most Christian and Jewish sects happily do without. (In recent times, Christian fundamentalism has been a peculiarly American movement; nowhere in Western Europe, in fact, has creationism been an issue.)

Second, most fundamentalists identify with an ultraconservative political movement. They long for the return of a more moral America, an America that may never have been. All around them they see what they perceive as declining morality and spirituality, as evidenced by pornography, crime, drug abuse, abortion, feminism, atheism, liberalism, and all sorts of other threatening isms. They reason that if humans share ancestry with the other animals, we have no reason to behave as anything other than animals. This view neglects the fact that humans are the only known animals with the ability to contemplate the consequences of their own actions. It also fails to recognize that there is a great deal of good in the world, the nightly news notwithstanding. Crime existed long before the theory of evolution, even before the writing of the Bible, and biologists do not like crime any more than the creationists do. Evolutionary theory is not a license to run amok, and neither is a belief in the literal interpretation of the Bible a guarantor of moral behavior.

How will all of this end? I suspect there will be an evolutionary ending. As future generations acquire more knowledge through education, simplistic answers based on belief will become increasingly unsatisfying. Those religions that cannot reconcile their beliefs with advancing scientific knowledge and common sense will lose followers to the more flexible, less dogmatic religions. Religions, after all, are themselves subject to evolution. The religions that are unable to adapt will leave no offspring, so to speak, and eventually will become extinct.

History offers us a glimpse of this process in action. In 1543 Copernicus argued that the Earth and planets revolve around the Sun. To the Catholic Church, which for centuries had insisted that the Earth was the center of everything, this was a great shock and a blasphemy. Copernicus's disciple, Bruno, was burned alive at the stake in 1600 for this and other heresies, just twenty years before the first settlers—leaving Europe to escape religious persecution and dogma—arrived in Massachusetts. Galileo (1564–1642) spent the last eight years of his life under house arrest for his support of the Copernican system. But today I know of no religion that teaches that the Earth is the center of the solar system, or that it is flat, although there are still geocentrists and flat-earthers who claim that their positions are supported by a literal interpretation of the Bible.

In 1859 Darwin completed the Copernican revolution by removing humans from center stage. By 1900, most of the scientific world was convinced of the validity of the theory of evolution. It is just a matter of time before this fruitful concept comes to be accepted by the public as wholeheartedly as it has accepted the spherical Earth and the Sun-centered solar system.

Appendixes

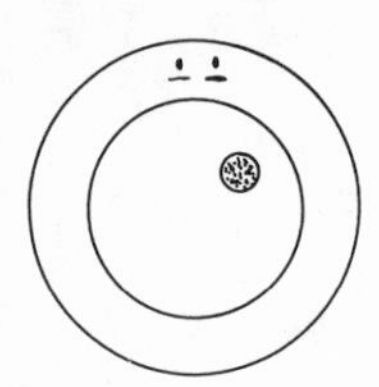

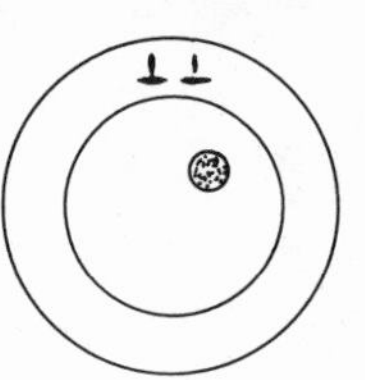

Interphase
Nuclear membrane complete;
Nucleolus visible;
Chromosomes not distinct.

Prophase (Early)
Centrioles moving apart;
Chromosomes appear as elongated threads;
Nucleus becoming less distinct.

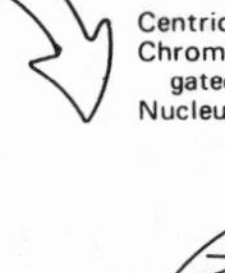

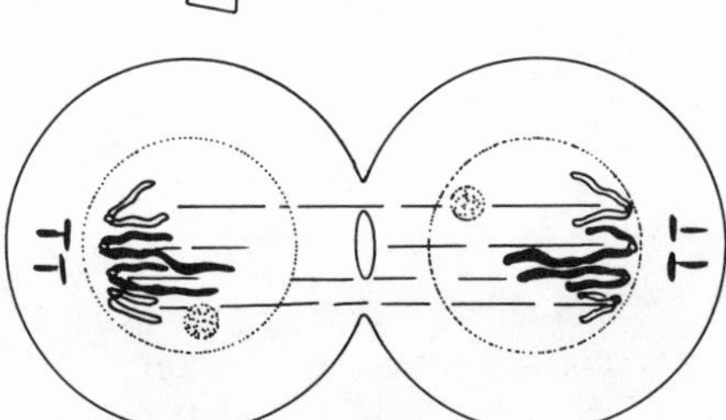

Telophase
Chromosomes becoming elongate, less distinct;
New nuclear membrane forming;
Nucleus replicated;
(Cell division nearly complete).

Prophase (Late)
Each chromosome composed of two chromatids attached by centromere;
Nucleolus no longer visible;
Centrioles at opposite ends of nucleus;
Spindle formed;
Nuclear membrane disintegrating;
Chromosomes moving toward equator of spindle.

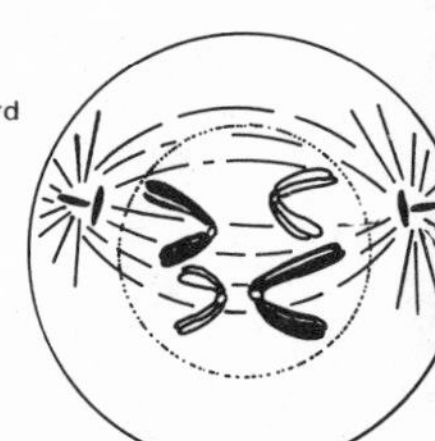

Anaphase
The two sets of chromosomes near their respective poles;
(Cell division beginning).

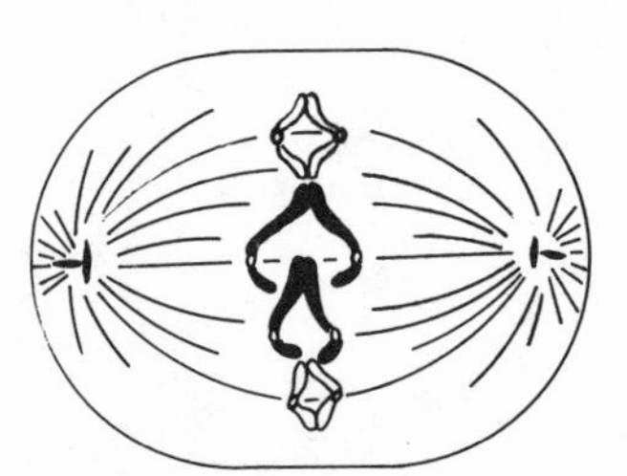

Metaphase

Centromeres divide and move toward opposite poles of spindle, dragging single stranded chromosomes with them.

Appendix A Chromosomes, Genes, and Genetic Variation

Before we can discuss the sources of heritable variation, we must take a brief tour of a cell to get acquainted with genetic material. All organisms are composed of **cells**. Some organisms, such as amoebas, consist of only one cell; others, such as vertebrates, are composed of trillions of cells. Within each cell is a **nucleus**, and within the nucleus are the **chromosomes**. Chromosomes are threadlike strands of **deoxyribonucleic acid** (**DNA**) and **protein**, and each chromosome carries numerous hereditary instructions encoded in its DNA.

Each animal and plant species has a characteristic number of chromosomes, and each cell in the body of an individual organism has the number of chromosomes specific to that species. As the body grows, as old cells die, or as wounds heal, new cells are formed in a process called cell division, or **mitosis**. During this process the chromosomes duplicate, and each new cell therefore continues to have the proper number of chromosomes (Figure A1). Cells with the complete chromosome complement are said to be **diploid**.

Figure A1 (facing page). Mitosis and cell division in an animal cell with two pairs of chromosomes. Mitosis is an equational division, which means that each daughter cell receives the same number of chromosomes as the parent cell.

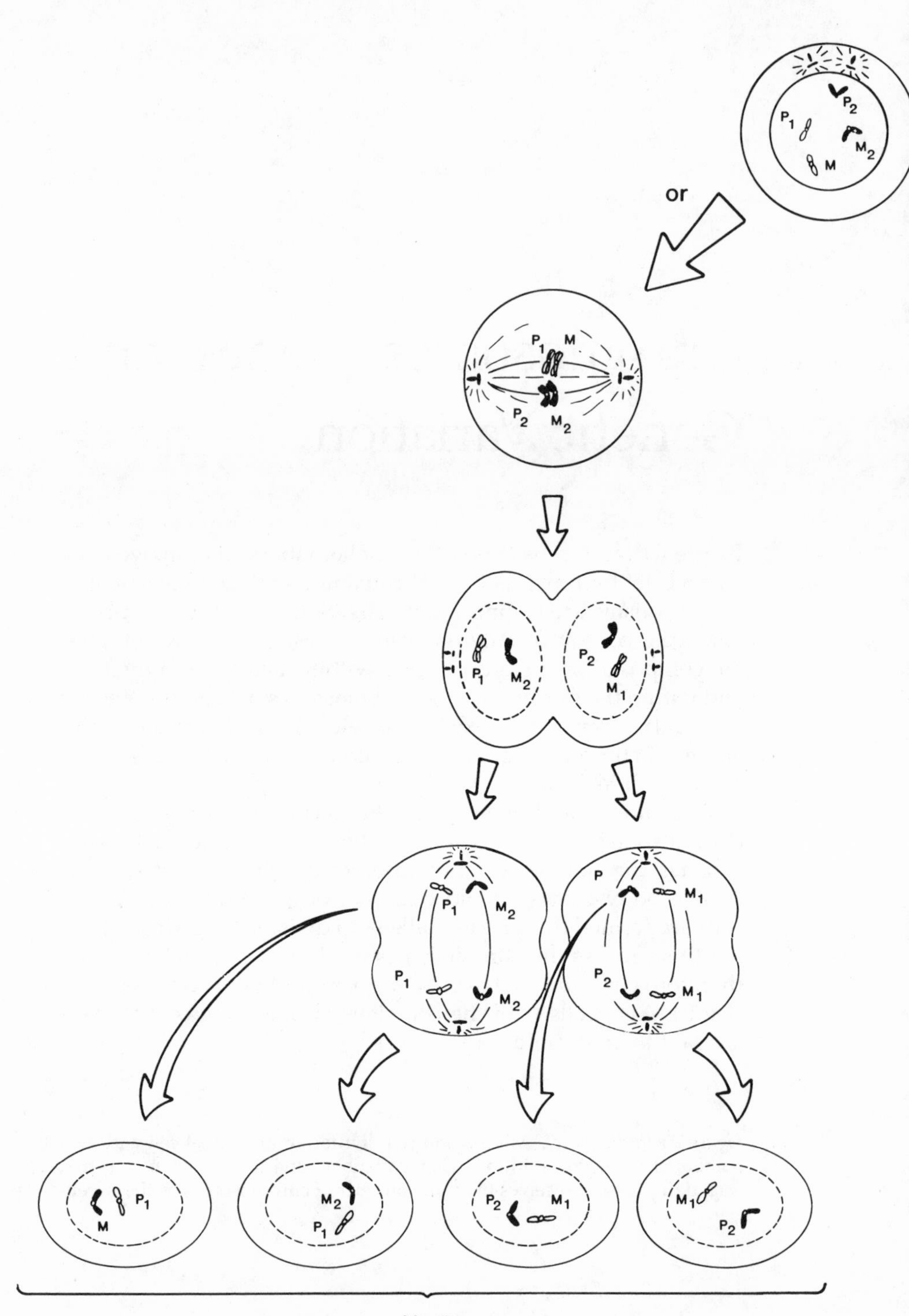

P2
P1
M2
M
or
P1
M
P2
M2
P1
M2
P2
M1
P
P1
M2
M1
P1
M2
P2
M1
M1
M
P1
M2
P1
P2
M1
M1
P2
N=2

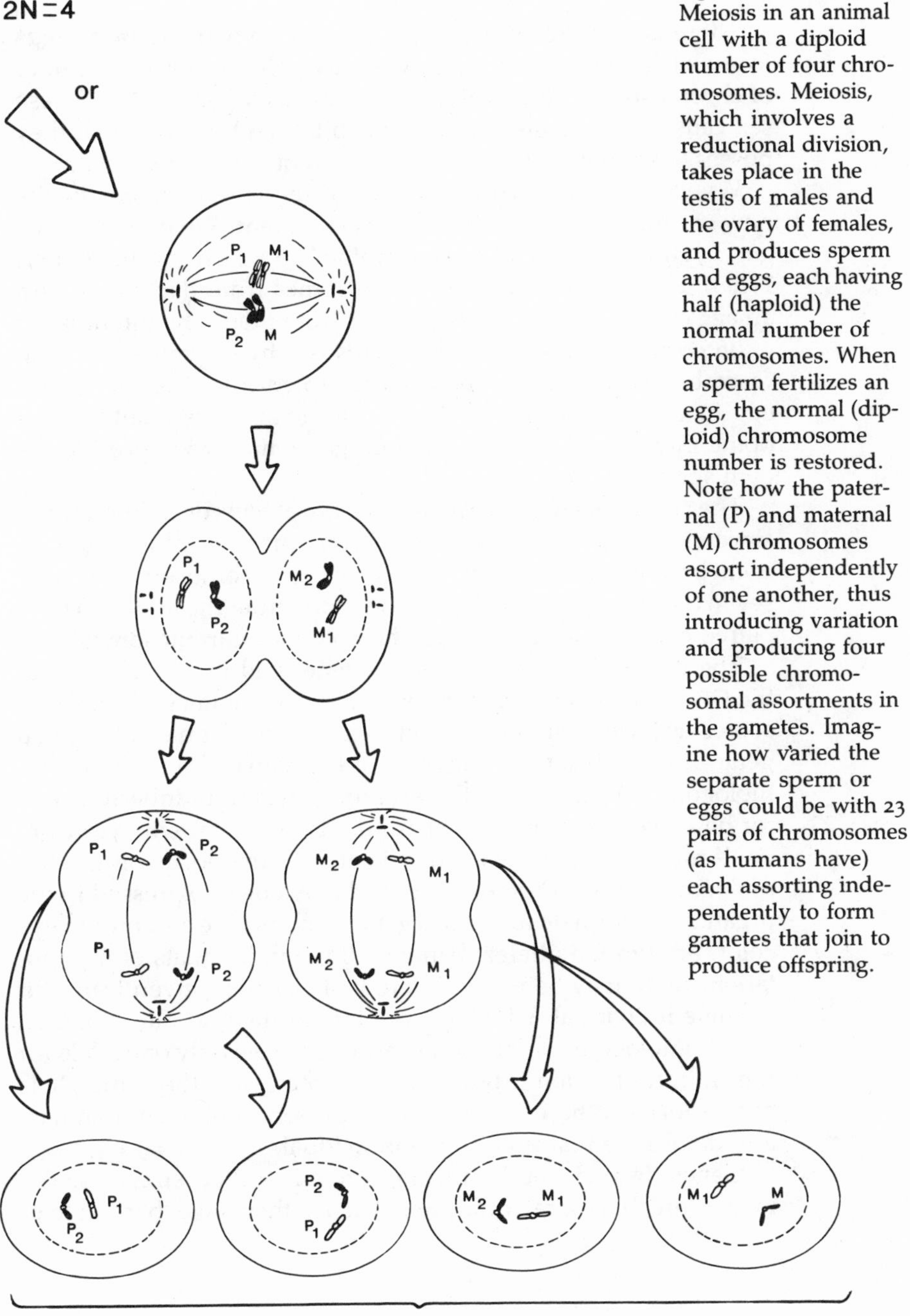

Figure A2 (both pages). Meiosis in an animal cell with a diploid number of four chromosomes. Meiosis, which involves a reductional division, takes place in the testis of males and the ovary of females, and produces sperm and eggs, each having half (haploid) the normal number of chromosomes. When a sperm fertilizes an egg, the normal (diploid) chromosome number is restored. Note how the paternal (P) and maternal (M) chromosomes assort independently of one another, thus introducing variation and producing four possible chromosomal assortments in the gametes. Imagine how varied the separate sperm or eggs could be with 23 pairs of chromosomes (as humans have) each assorting independently to form gametes that join to produce offspring.

There is one exception to this process. When **sperm** or **eggs** form, each sperm or egg receives only half the normal number of chromosomes, in a process called **meiosis** (Figure A2). These sex cells, called **gametes**, are said to be **haploid**, because they possess only half of the normal amount of genetic material.

When sperm fertilizes egg, half of the chromosomes in the new organism are inherited from each parent. We humans each get 23 chromosomes from our mother and 23 from our father. When a sperm cell with 23 chromosomes from dad fuses with an egg cell carrying 23 chromosomes from mom, the full number of 46 chromosomes is restored. This is why organisms tend to resemble both parents. The **zygote** formed from the sperm and egg fusion then begins to divide, by mitosis, eventually developing into an adult capable of undergoing meiosis to produce its own sperm or egg.

The chromosomes in a diploid cell are always even in number, since chromosomes come in pairs, one member of a pair from each parent. For example, one copy of chromosome number seven comes from the male, and one copy from the female. These paired chromosomes are called **homologous chromosomes**.

The unit of heredity is called a **gene**, and it is a segment of the DNA molecule along a chromosome. If you imagine chromosomes as strings of genes, then there are two copies of a given gene—one each at the same place or **locus** on each of the two homologous chromosomes. For example, one might inherit a gene for blue eyes from one's father and a gene for brown eyes from one's mother. These two genes lie at the same locus on the homologous pair of chromosomes, and only one is expressed in the organism. Alternate forms of a gene, such as blue and brown eye color, are termed **alleles**. Among all the individuals of a population, there may be many different alleles for a given trait, but a single individual will have only two of them at most (one on each homologous chromosome) and may have only one allele for the trait in question if both chromosomes have the same allele at that locus. Where two alleles are present, one may dominate and determine that trait in that individual.

Genes do their job by making proteins called **enzymes**. Enzymes are biological **catalysts**; that is, they take part in bio-

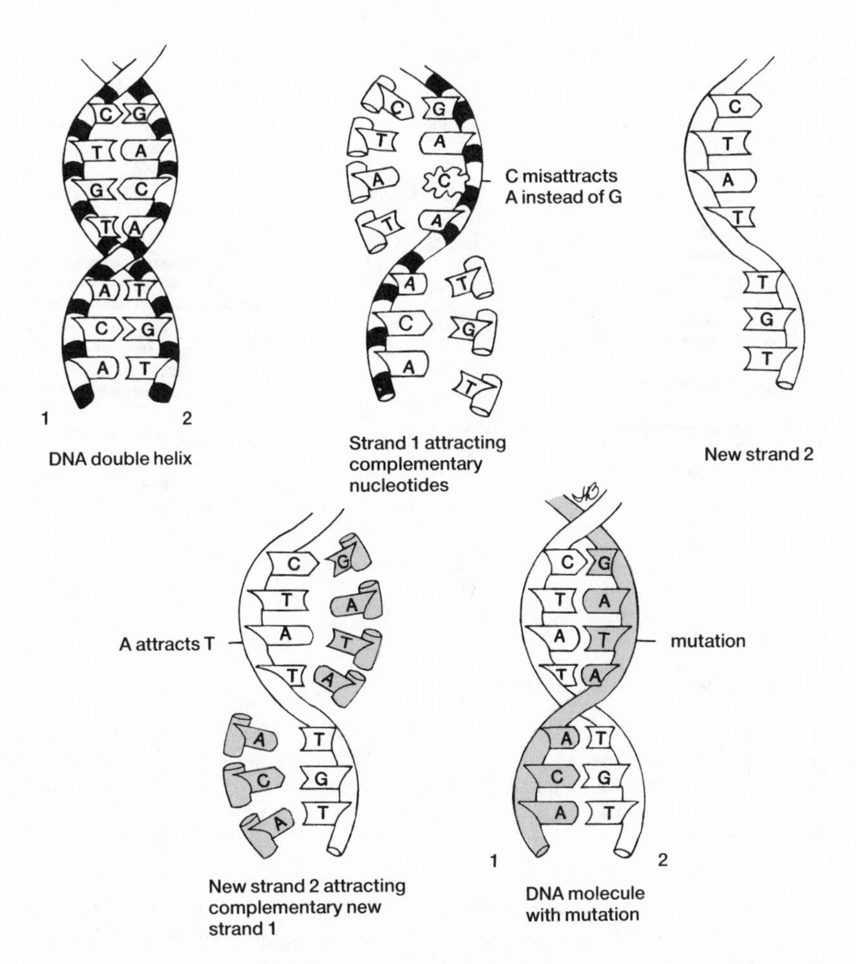

Figure A3. Replication and mutation. The DNA molecule (left) of any living organism is a double helix. The two strands (here, of just one small section of the full DNA molecule) are held together by the nitrogenous base pairs, just as the steps of a ladder hold the rails. There are four bases: adenine (A), thymine (T), cytosine (C), and guanine (G). A joins with T, and C with G. The rails of the DNA ladder are composed of a deoxyribose sugar (unshaded) and a phosphate group (black); a unit of sugar, phosphate, and base is called a nucleotide. During cell division the two DNA strands separate. Each strand attracts complementary nucleotides to reconstruct the double helix. In the second figure an abnormal cytosine misattracts adenine instead of its usual guanine. When the new strand 2 forms its new complementary strand (fourth figure), adenine attracts thymine, as it should. The new DNA molecule now has an A-T pair instead of the G-C pair of the grandparent DNA (last figure). This mutation affects the code that determines protein formation, as shown in Figures A4 and A5.

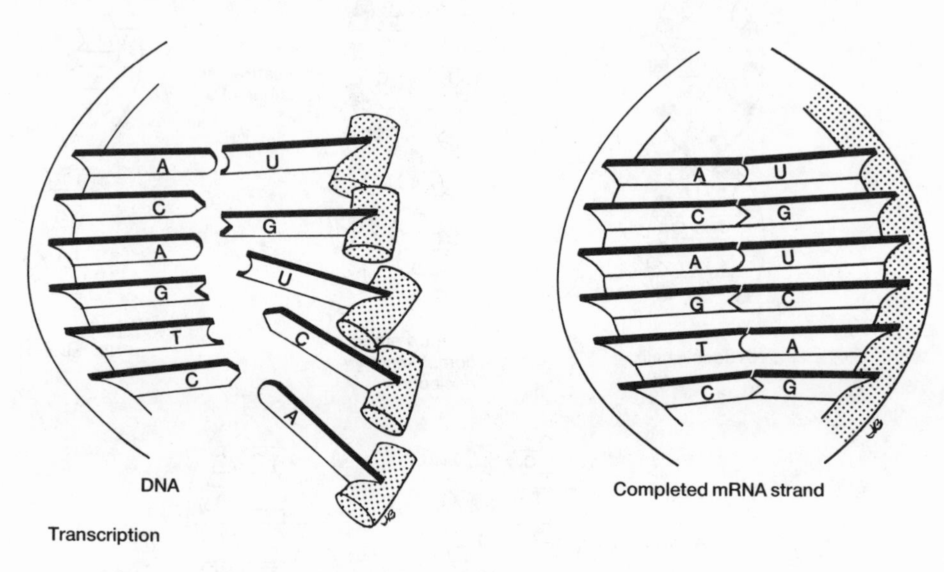

Figure A4. Transcription, the formation of messenger RNA (ribonucleic acid) from the coded DNA. The DNA molecule unzips and serves as a template for mRNA synthesis. The mRNA bases are complements of the DNA bases even though uracil (U) replaces thymine (T).

chemical reactions, but are not used up in those reactions. They facilitate specific chemical transformations that control all the reactions in the cells, and thereby in the organisms the cells compose. Each gene controls the manufacture of a particular enzyme. Figures A3–A5 and Table A1 detail the process whereby these proteins are made from the instructions encoded by the genes.

With this brief background in mind, we can begin to explore where genetic variation comes from. If there were no genetic change, there would be no evolution, since every offspring would be identical to its parents. But genes do change. The sources of heritable variation in nature include **mutation**, chromosomal rearrangement, and independent assortment of genes from different chromosomes. These sources combined with the randomness of sexual reproduction, ensure the presence of genetic variation.

A mutation is a change in a gene, occurring during the lifetime of an individual organism and produced by some agent in the external environment. Mutations may be brought about by ionizing radiation, chemicals, heat, or other means. Mutations create new alleles, whereas the other processes sort existing alleles into new combinations in new individuals. The complicated chemical

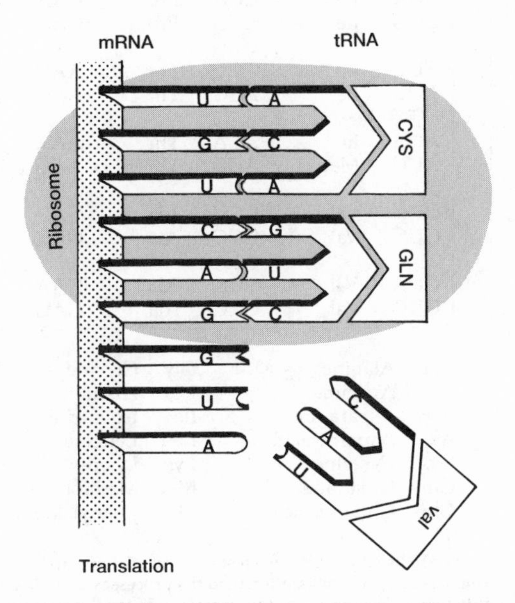

Figure A5. Translation, the process in which the genetic information coded in mRNA (from DNA) directs the specific amino acid sequence during protein synthesis. The mRNA moves out of the cell nucleus into the cytoplasm, where it attaches to cell organelles that function in protein synthesis (ribosomes). Another kind of RNA called transfer RNA (tRNA) picks up free amino acids within the cell. The sequence of three bases on the tRNA (anticodon) is attracted to the complementary sequence (codon) on the mRNA. This causes the amino acids carried by the tRNA to become ordered in a given sequence as the ribosome moves along the mRNA. The ordering of the amino acids results in the formation of a specific protein under the direction of the sequence of codons in DNA, as transcribed and translated by mRNA and tRNA, respectively.

TABLE A1

Genetic Code Specifying Amino Acid Sequence

First Base	Second Base: U		C		A		G		Third Base
U	UUU	Phe	UCU	Ser	UAU	Tyr	UGU	Cys	U
U	UUC	Phe	UCC	Ser	UAC	Tyr	UGC	Cys	C
U	UUA	Leu	UCA	Ser	UAA	*Stop*	UGA	*Stop*	A
U	UUG	Leu	UCG	Ser	UAG	*Stop*	UGG	Trp	G
C	CUU	Leu	CCU	Pro	CAU	His	CGU	Arg	U
C	CUC	Leu	CCC	Pro	CAC	His	CGC	Arg	C
C	CUA	Leu	CCA	Pro	CAA	Gln	CGA	Arg	A
C	CUG	Leu	CCG	Pro	CAG	Gln	CGG	Arg	G
A	AUU	Ile	ACU	Thr	AAU	Asn	AGU	Ser	U
A	AUC	Ile	ACC	Thr	AAC	Asn	AGC	Ser	C
A	AUA	Ile	ACA	Thr	AAA	Lys	AGA	Arg	A
A	AUG	Met	ACG	Thr	AAG	Lys	AGG	Arg	G
G	GUU	Val	GCU	Ala	GAU	Asp	GGU	Gly	U
G	GUC	Val	GCC	Ala	GAC	Asp	GGC	Gly	C
G	GUA	Val	GCA	Ala	GAA	Glu	GGA	Gly	A
G	GUG	Val	GCG	Ala	GAG	Glu	GGG	Gly	G

Ala	Alanine	Gly	Glycine	Pro	Proline
Arg	Arginine	His	Histidine	Ser	Serine
Asn	Asparagine	Ile	Isoleucine	Thr	Threonine
Asp	Aspartic acid	Leu	Leucine	Trp	Tryptophan
Cys	Cysteine	Lys	Lysine	Tyr	Tyrosine
Gln	Gluamine	Met	Methionine	Val	Valine
Glu	Glutamic acid	Phe	Phenylalanine		

NOTE: To use this table, which shows the relationship between messenger RNA and specific amino acids, one must understand the processes described in Figures A3–5. The DNA molecule (Figure A3) provides the genetic code for the formation of mRNA (Figure A4), which, in turn, directs protein synthesis by ordering specific amino acid sequences (Figure A5). The table shows the code for each amino acid of which proteins are composed. The coding unit, or codon, is a sequence of three bases from the mRNA molecule. Note that most of the amino acids are specified by several different codons, many of which differ only in the third base. Thus there are 64 possible codons (4^3) that specify only 20 amino acids and three stop signs (which halt protein synthesis). This redundancy allows a great deal of genetic change that is not reflected in the protein and is, consequently, not subject to natural selection. For example, the amino acid leucine is specified by six different codons. If UUA mutates to CUG, no difference will be apparent, because leucine is still formed and the protein will be unchanged. To find the amino acid specified by a particular codon, locate the first base at the left, then go across to the column headed by the second base, then go down the right side to locate the third base.

details of mutational change are explained in Figures A3–A5 and Table A1.

The same forces that produce mutations bring on chromosomal rearrangements. These rearrangements may result in a reversal of the order of genes in a segment of a chromosome. Such a change is called an **inversion**. A **deletion** occurs if a segment of a chromosome, with its attendant genes, is lost. Occasionally, chromosomes will exchange segments; this "crossing over" is called a **translocation**. If a piece of a chromosome joins another chromosome, it is called **fusion**. All of these highly complicated changes—inversion, deletion, translocation, and fusion—alter the order of genes in the chromosome and provide important sources of variability.

During sperm and egg formation, each gamete receives a haploid set of chromosomes. But which chromosome from each homologous pair is transmitted to a given gamete is random and independent of the other chromosomes it receives. This is most easily understood by looking at the illustration of gamete formation (Figure A2).

Sexual reproduction combines a sperm and an egg (each produced by meiosis with its attendant shuffling of chromosomes) into a new individual—a unique combination of genes. Sexual reproduction is in fact the major source of genetic variability.

Appendix B A Chronology of Charles Darwin

Date	Event	Darwin's Age
1809 February 12	Born at Shrewsbury, son of Robert Waring Darwin and Suzannah née Wedgwood	
1817	Attended Dr. G. Case's day school at Shrewsbury	8
1818	Entered Shrewsbury School	9
1825 October 22	Matriculated at University of Edinburgh	16
1827 October 15	Admitted to Christ's College, Cambridge	18
1831 April 26	Graduated with B.A. degree	22
1831 August 30	Received invitation to sail on the *Beagle*	22
1831 December 27	H.M.S. *Beagle* sailed from Plymouth	22
1835 March 26	Heavily bitten by *Triatoma infestans* in the Andes	26
1835 September	Visited Galapagos Islands	26
1836 October 2	Landed at Falmouth, England	27
1837 March 13	Took quarters at 36 Great Marlborough Street, London	28
1837 July	Opened first notebook on transmutation of species	28

Date	Event	Darwin's Age
1838 September 28	Read Malthus on *Population*	29
1838 November 11	Proposed marriage to Emma Wedgwood, his cousin, and was accepted	29
1839 January 1	Established at 12 Upper Gower Street, London	29
1839 January 24	Elected Fellow of the Royal Society	29
1839 January 29	Married Emma Wedgwood at Maer, Staffordshire	29
1839 August	*Journal of Researches into the Geology and Natural History* (of the voyage of the *Beagle*) published	30
1839 December 27	William Erasmus Darwin born	30
1841 March 1	Anne Elizabeth Darwin born	32
1842 May	Wrote sketch of species theory; *Structure and Distribution of Coral Reefs* published	33
1842 September 17	Moved to Down House, Downe, Kent	33
1842 September 23	Mary Eleanor Darwin born (died three weeks later)	33
1843 September 25	Henrietta Emma Darwin born	34
1844 July	Wrote essay on species theory	35
1844 November	*Geological Observations on Volcanic Islands* published	35
1845 July 9	George Howard Darwin born	36
1845 August	*Journal of Researches*, 2nd edition, published	36
1846 October 1	Began work on barnacles; *Geological Observations on South America* published	37
1847 July	Elizabeth Darwin born	38
1848 August 16	Francis Darwin born	39
1850 January 15	Leonard Darwin born	40
1851 April 23	Anne Elizabeth Darwin died	41
1851 May 13	Horace Darwin born	41

Date	Event	Darwin's Age
1851 June	*Monograph of Fossil Lepadidae* published (barnacles); *Monograph of* [recent] *Lepadidae* published	42
1853	Awarded Royal Medal of the Royal Society	44
1854 August	*Monograph of* [recent] *Balanidae* published; *Monograph of Fossil Balanidae* published	45
1854 September 9	Began to sort out notes on species	45
1856 May 14	Began to write large work on species	47
1856 December 6	Charles Waring Darwin born	47
1858 June 18	Received from Alfred Russel Wallace a perfect summary of his own theory of evolution by natural selection	49
1858 June 28	Charles Waring Darwin died	49
1858 July 1	Joint paper with Wallace read before Linnean Society of London	49
1858 July 20	Began to write *On the Origin of Species*	49
1858 August 20	Joint paper with Wallace published	49
1859 November 24	*On the Origin of Species* published (1,250 copies, all sold on first day)	50
1860 January 7	*Origin* (2nd edition) published (3,000 copies)	51
1861 April	*Origin* (3rd edition) published (2,000 copies)	52
1862 March 1	Paper on dimorphism in *Primula* published	53
1862 May 15	*On the Various Contrivances by which British and Foreign Orchids are Fertilised by Insects* published	53

Date	Event	Darwin's Age
1864 November 30	Awarded Copley Medal of the Royal Society	55
1866 December 15	*Origin* (4th edition) published (1,250 copies)	57
1868 January 30	*Variation of Animals and Plants under Domestication* published (1,500 copies)	58
1868 February 20	*Variation of Animals and Plants under Domestication*, reprint published (1,500 copies)	59
1869 August 7	*Origin* (5th edition) published (2,000 copies)	60
1871 February 24	*Descent of Man* published (2,500 copies and reprint of 5,000)	62
1872 February 19	*Origin* (6th edition) published (3,000 copies)	63
1872 November 26	*The Expression of the Emotions in Man and Animals* published (7,000 copies and reprint of 2,000)	63
1874 June	*Structure and Distribution of Coral Reefs* (2nd edition) published	65
1874 Autumn	*Descent of Man* (2nd edition) published	65
1875 July 2	*Insectivorous Plants* published	66
1875 September	*Climbing Plants* published	66
1875 End	*Variation of Animals and Plants Under Domestication* (2nd edition) published	66
1876 December 5	*Effects of Cross- and Self-Fertilisation in the Vegetable Kingdom* published	67
1877 January	*Fertilisation of Orchids* (2nd edition) published	68
1877 July 9	*Different Forms of Flowers on Plants of the Same Species* published	68
1879 November 19	*Life of Erasmus Darwin* published	70

Date	Event	Darwin's Age
1880 November 22	*Power of Movement in Plants* published	71
1881 October 10	*Formation of Vegetable Mould Through the Action of Worms* published	72
1882 April 19	Died at Down House	73
1882 April 26	Buried in Westminster Abbey	

Modified from Gavin de Beer (ed.). 1974. *Charles Darwin/Thomas Henry Huxley Autobiographies*. Oxford Univ. Press, London. 123 pp. And from: Gavin de Beer. 1965. *Charles Darwin: A Scientific Biography*. Anchor Books, Garden City, N.Y. 295 pp. Courtesy of Oxford University Press.

Reference Material

Glossary

Most technical terms used in this book, especially those given in text in **boldface** type, are defined in this Glossary. Most technical terms employed within the definitions in this Glossary are also defined here. Terms are liberally cross-referenced to related terms.

absolute age. The precise geologic age of a fossil or rock, usually determined by radiometric dating. *Compare* relative age.

adaptation. An anatomical structure, physiological process, or behavioral trait that enhances an individual's survival and reproductive success in a given environment; the process of evolutionary modification of an organism to better suit its environment, which results in improved survival and reproductive success. *See* natural selection, artificial selection.

adaptive radiation. The evolutionary divergence of members of a single phylogenetic lineage into a variety of ecological roles, usually over a relatively short period of time and resulting in the appearance of several or many new species.

aerobic respiration. The exchange of gases between an organism and its environment whereby oxygen is taken in and carbon dioxide is given off. *See* mitochondrion.

albinism. The absence of pigmentation in skin, hair, or eyes, owing to a recessive gene. *Compare* melanism.

allele. One of two or more different forms of a gene coding for the same trait and located at the same locus or position on homologous chromosomes, as for example blue eyes and brown eyes.

allopolyploidy. The phenomenon of hybridization based on a doubling of the chromosomes of an organism, such that it contains chromosome sets from two or more unlike sources (different species or even genera). *Compare* polyploidy.

amber. A yellow or reddish-brown, translucent to transparent fossil resin exuded by ancient trees, often containing well-preserved insects or other organisms.

amino acid. One of the 20 naturally occurring building blocks of protein, each an organic compound with an amino group (—NH_2) and an acid group (—COOH).

amniotic egg. The shelled egg of reptiles, birds, and mammals, in which a membrane forms a fluid-filled sac around the embryo, thus enabling the animal's life cycle to be completed on land.

anatomy. The scientific study of the structure of organisms. *Compare* morphology, physiology.

anthropology. The scientific study of human evolution, physical variability, culture, and behavior, past and present. *Compare* archaeology, primatology, paleoanthropology.

antibody. A protein produced as a defense mechanism by a lymphocyte, usually circulating in the blood, that is able to attack (combine with and neutralize) the antigen that stimulated its production. *See* immune response, antigen.

antigen. Any foreign substance, most often a protein, capable of stimulating antibody formation (the immune response) in an organism, for example a virus or plant pollen. *See* antibody.

arboreal. Adapted for living in trees, as a monkey or squirrel. *Compare* terrestrial.

archaeology. The scientific study of prehistoric human culture by means of the remains left by ancient peoples. *Compare* anthropology, primatology, paleoanthropology, paleontology.

artificial selection. Human control of the gene pool of a population, as in the selective breeding of domestic animals or horticultural plants. *Compare* natural selection; see adaptation.

asexual reproduction. Production of new individuals that are genetically identical to the parent, by means that do not involve fusion of sperm and egg, such as by budding or division into two or more parts. *Compare* sexual reproduction, clone.

atomic nucleus. *See* nucleus.

balanced polymorphism. Maintenance of equilibrium in the relative proportions of several forms in a population, resulting when heterozygotes are favored by selection over both homozygous types.

behavioral isolation. Reproductive isolation owing to inappropriate courtship behavior and, in consequence, to the inhibition of mating.

Big Bang. The cosmological theory that all the matter and energy in the Universe originated in a superdense agglomeration that exploded at a finite instant in the past. *See* cosmology.

bilateral symmetry. The arrangement of body parts wherein an animal has a left and a right side, a front and a rear, and back and belly sides. *Compare* radial symmetry.

biochemistry. The study of the chemistry of living organisms. *Compare* organic chemistry, metabolism.

biogeography. The study of the geographical distribution of plants and animals, past or present.

biological control. The control of pests by biological rather than chemical means, as for example by introducing ladybird beetles into an infestation of aphids.

biology. The scientific study of living (or once living) organisms and vital processes.

bipedality. Walking erect on the hind limbs.

brow ridge. The prominent ridge of skull bone arching over each eye and often connected above the bridge of the nose.

budding. A form of asexual reproduction in which a new individual is produced as an outgrowth of the parent, as found in yeast, *Hydra*, and sponges. *See* asexual reproduction.

catalyst. A substance that increases the rate of a particular chemical reaction without itself undergoing any permanent chemical change.

cell. The structural and functional unit of all living organisms.

cell organelle. *See* organelle.

chemistry. The study of the elements of matter and of the molecules and compounds they form. *See* biochemistry, prebiotic chemistry.

chlorophyll. The green, light-absorbing pigments of plants necessary for photosynthesis.

chloroplast. A cell organelle that houses chlorophyll and is the site of photosynthesis.

chromosome. A threadlike body in the nucleus of an animal or plant cell that contains the genes, composed of DNA and protein, by which hereditary information is transmitted from one generation to the next.

chromosome segment. A portion of a chromosome with some of its genes.

cline. More or less continuous variation in a character of the individuals of a species, for example in shell color, along an environmental gradient such as altitude or latitude, produced by changing proportions in the gene pool in response to selection.

clone. A group of cells or organisms derived asexually from a single individual and therefore genetically identical; one of two or more such identical cells or organisms, as for example plantations of identical trees all derived from the tissue of a single plant. *See* asexual reproduction.

coacervate. An aggregation of lipid molecules in water, held in suspension by electrostatic forces; mayonnaise is a coacervate. *Compare* prebiotic chemistry.

coagulate. To clot; to change from a fluid to a congealed mass.

codon. The basic coding unit of a DNA molecule, consisting of three nucleotides that determine the synthesis of a specific amino acid. *See* Table A1 in the Appendix.

colony. *See* population.

commensalism. The living together of unlike organisms such that one (the commensal) benefits from the association while the other, the host, neither benefits nor suffers from it. *Compare* symbiosis, mutualism, parasitism.

compound. A substance formed by the chemical combination of two or more elements. *See* molecule.

convergence. *See* evolutionary convergence.

coprolite. Fossilized feces.

cosmology. The scientific study of the origin, structure, and future of the Universe. *See* Big Bang.

cytochrome. One of several iron-containing pigments that serve as electron carriers in aerobic respiration in cells and organisms.

decay. *See* radioactive decay.

deduction. The process of deriving a conclusion by reasoning from particular cases to general principles.

deletion. A mutation in which a central section of a chromosome, with its attendant genes, is lost during mitosis or meiosis.

deoxyribonucleic acid (DNA). The genetic material of organisms, a major component of chromosomes in the cell nucleus, organized into a linear series of genes that determine hereditary characteristics by controlling protein synthesis. DNA is a double-stranded, helical molecule composed of long chains of sugar and phosphate groups with four nucleotide bases (adenine, thyamine, cytocine, guanine) as side groups.

derived. Having advanced characteristics not found in the ancestral stock. *Compare* primitive.

differential reproduction. Reproduction wherein one form of a variation leaves more offspring than another form of the variation; natural selection. *Compare* adaptation.

dimorphism. Having two very different forms, as for example a tadpole and an adult frog. *See* sexual dimorphism.

diploid. Possessing a double set of homologous chromosomes, which is the normal number of chromosomes present in the body cells (but not the sex cells) of sexually reproducing organisms; $2n$; two times the haploid number.

divergence. *See* evolutionary divergence.

DNA. *See* deoxyribonucleic acid.

DNA-DNA hybridization. A method of determining the similarity of DNA from different organisms, and thus the presumed degree of relatedness of the organisms. *Compare* molecular clock.

dominance. The ability of one allele to mask the expression of its partner at the same locus on the homologous chromosome in a diploid cell.

ecological niche. The functional role of a plant or animal in its community. *Compare* habitat, environment.

egg. The mature, usually immobile, female gamete of sexually reproducing organisms. *Compare* sperm.

embryo. The life stage of a plant or animal beginning from the zygote and terminating with germination (in plants) or hatching or birth (in animals).

embryology. The study of the growth and early development of an organism.

environment. An organism's surroundings, including other organisms as well as climate, soil, water, temperature, landform, and other nonliving things. *Compare* habitat, ecological niche.

environmental gradient. A regularly increasing or decreasing change in a parameter such as rainfall, temperature, altitude, or latitude. *See* cline.

enzyme. A protein that acts as a biological catalyst.

equilibrium. A balance between two or more opposing forces. *Compare* stasis.

eukaryote. An organism each cell of which contains a discrete membrane-bound nucleus, as is characteristic of all organisms except bacteria and cyanobacteria. *Compare* prokaryote.

evolution. Descent with modification; a change in gene frequency. *See* macroevolution, microevolution.

evolutionary convergence. The development of similar characteristics by unrelated organisms, in response to common environmental pressures, as in sharks and porpoises.

evolutionary divergence. The development of different characteristics by related organisms, as in dogs and seals.

experiment. A controlled procedure carried out to discover or test something. *See* scientific method.

extinction. The elimination of the last individual of a species or of a whole lineage of organisms. *Compare* prehistoric.

fauna. All the animals of a given region and/or time period. *Compare* flora.

faunal succession. The steady replacement of the characteristic animals of a fauna by others through time in a definite and recognizable sequence.

fitness. The measure of an organism's success in getting its genes into the next generation, owing to its adaptation to its environment.

flora. All the plants of a given region and/or time period. *Compare* fauna.

floral succession. The steady replacement of the characteristic plants of a flora by others through time in a definite and recognizable sequence.

fossil. Any remains or traces of prehistoric life. *See* indicator fossil, transitional fossil.

founder effect. The principle that only a small amount of the total genetic variation in a source population is carried by the few founding individuals of a new population.

galaxy. A huge collection of stars, gases, and dust held together by mutual gravitational attraction; the Milky Way is our galaxy.

gamete. One of the two haploid reproductive cells, the sperm of the male or the egg of the female.

gametic isolation. Reproductive isolation owing to the inability of sperm and egg to reach one another.

gene. The basic unit of inheritance on a chromosome, composed of a specific sequence of nucleotides in a DNA molecule that codes for a single polypeptide.

gene flow. The exchange of genes within a population or between populations by interbreeding or migration.

gene frequency. The proportion of a given allele, relative to the total of all alleles at the same locus, in a population of a species.

gene pool. The total of all the alleles of all genes in a population.

genetic code. The sequence of nucleotide base pairs in the DNA molecule that carries hereditary information by directing the formation of proteins.

genetic drift. Random change in allele frequencies over time in a population or species, owing to chance occurrence rather than to natural selection.

genetic engineering. Alteration of an organism's DNA by the artificial insertion of genes from another organism.

genetic variation. Inheritable divergence of structural or functional characteristics within a population.

genetics. The study of heredity and variation in organisms and populations.

genome. The total of all the genes in a haploid set of chromosomes in an individual organism.

geochemistry. The scientific study of the chemical composition of the Earth.

geological formation. A body of rock strata united by certain shared physical characteristics.

geology. The scientific study of the origin, structure, and composition of the Earth.

geophysics. The scientific study of the physics of the Earth, including the fields of meteorology, hydrology, seismology, volcanology, and others.

germ cell. A sex cell or gamete; a sperm or egg.

gradualism. The mode of evolution in which species change by relatively slow, stepwise transformations through time. *Compare* punctuated equilibrium.

habitat. The specific location where an organism generally lives, as for example a pond, for a frog. *Compare* ecological niche, environment.

half-life. The time required for the decay of half of the original parent isotope in a radioactive sample, the basis for radiometric dating.

haploid. Possessing a single set of unpaired chromosomes, which is the normal number of chromosomes for gametes (sex cells); half the number of chromosomes found in a somatic cell; *n*; half the diploid number.

hemoglobin. The iron-containing protein of vertebrate (and some invertebrate) blood that transports oxygen.

heredity. The transmission of genetic characters from parents to offspring.

hermaphrodite. An organism that possesses both male and female sex organs in the same body.

herpetology. The scientific study of amphibians and reptiles.

heterozygote. A diploid organism that possesses two different alleles for any one gene. *Compare* homozygote.

hominid. A primate of the human family (Hominidae), which includes *Australopithecus* and *Homo*.

hominoid. A primate of the gibbon family (Hylobatidae), ape family (Pongidae), or human family (Hominidae), the three taken together as the superfamily Hominoidea.

homologous chromosome. One of two chromosomes of a pair that are identical to one another with respect to gene loci but are derived from different parents.

homozygote. A diploid organism that possesses two identical alleles for any one gene. *Compare* heterozygote.

hybrid. The offspring of a reproductive cross between genetically dissimilar individuals, of for example two subspecies or species.

hybrid breakdown. The inability of the second generation of hybrids to survive or reproduce.

hybrid inviability. The inability of the first generation of hybrids to reach sexual maturity.

hybrid sterility. The inability of the first generation of hybrids to reproduce, owing to sterility.

hypothesis. A working explanation of processes in nature that leads to testable predictions, which may, when bolstered by facts and experimental results, develop into a theory. *Compare* theory, scientific method.

ichthyology. The scientific study of fishes.

immune response. The production of antibodies following exposure to an antigen; one of the body's defenses against infection.

immunity. The ability of a host organism to resist or overcome an infective agent.

immunology. The science that deals with the phenomena and causes of immunity, or resistance to disease or infection.

immutable. Unchangeable, unchanging.

indicator fossil. A fossil associated with a given geological stratum, and useful in identifying the relative age of similar strata elsewhere.

interbreeding. Mating between different individuals, populations, subspecies, or species.

inversion. A reversal of part of the normal linear arrangement of genes along a chromosome segment.

isolation. *See* reproductive isolation.

isotope. Any of two or more forms of a chemical element having in all of its atoms the same number of protons in the nucleus but a different number of neutrons.

lineage. The line of descent from a particular ancestor; a major group of plants or animals across a span of time, all members of which derive from a common ancestor.

lipid. One of a variety of fatlike organic compounds occurring in living organisms.

locus, pl. **loci.** The location of a gene (a segment of DNA) in the chromosome of an organism.

lymphocyte. A type of white blood cell that functions in the immune response.

macroevolution. Evolution above the species level; the collective major trends by which one or more lineages evolve through geologic time. *Compare* microevolution.

macromolecule. A very large molecule such as a protein or nucleic acid.

mammalogy. The scientific study of mammals.

mass number. The total number of neutrons and protons in an atomic nucleus.

mechanical isolation. Reproductive isolation owing to the inability of male and female genitalia to fit one another, thereby preventing copulation.

meiosis. A two-stage reductional cell division that results in gamete formation, each gamete having half the chromosome number of the parent cell. *Compare* mitosis; *see* Figure A2 in the Appendix.

melanism. Black coloration of the skin or hair brought about by excessive production of the pigment melanin under genetic control. A black squirrel is a melanistic gray squirrel, and a black panther is a melanistic leopard. *Compare* albinism.

metabolism. The sum of the biochemical reactions going on in a living organism by which energy is provided for vital processes, new material is assimilated, and old material is eliminated. *Compare* biochemistry.

microevolution. Evolution involving small changes in gene frequencies within a population over a few or many generations, up to the formation of new species. *Compare* macroevolution.

missing link. *See* transitional fossil.

mitochondrion, pl. **mitochondria.** The organelle responsible for aerobic respiration in the cell.

mitosis. Equational cell division that results in two new cells, each having the same chromosomes as the parent cells. It is the mechanism for growth, wound repair, and nonsexual reproduction. *Compare* meiosis; *see* Figure A1 in the Appendix.

molecular clock. The hypothesis that neutral mutations accumulate at a constant rate and therefore can be used to measure evolutionary divergence among species or lineages. *Compare* DNA-DNA hybridization.

molecule. The smallest structural unit that possesses the properties of a chemical compound. The molecule of water, for example, consists of two atoms of hydrogen and one of oxygen.

monogamy. A prolonged and more or less exclusive breeding relationship between a male and a female of a species. *Compare* pair bonding, polygamy.

morphology. The scientific study of the form and structure of organisms. *Compare* anatomy, physiology.

mutable. Subject to change.

mutation. A sudden heritable change in a gene; a change in the structure or amount of DNA in the chromosome induced by an environmental factor such as cosmic rays, heat, or chemicals.

mutation pressure. A process for change in gene frequency owing to the more frequent occurrence of a mutation (from A to B) than its back mutation to the original form (from B to A).

mutualism. The living together of unlike organisms such that both benefit from the association. *Compare* symbiosis, commensalism, parasitism.

natural selection. The mechanism by which the organisms best adapted to a given environment leave more offspring, thereby spreading the adaptation; differential reproduction. *Compare* adaptation, artificial selection.

neoteny. The acquisition of sexual maturity while retaining otherwise juvenile characteristics.

neutral mutation. A sudden genetic change that is neither favored nor disfavored by natural selection.

neutron. An uncharged elementary particle present in the nucleus of every atom that is approximately equal in mass to the proton. *See* mass number, nucleus, proton.

niche. *See* ecological niche.
notochord. A stiff, flexible rod that supports the embryo of chordates (vertebrates and their more primitive relatives); in adult vertebrates it is replaced by the vertebral column.
nuclear membrane. A lipid and protein layer surrounding the nucleus of eukaryotic cells.
nucleic acid. Either of two complex organic acids (DNA or RNA) composed of long chains of nucleotides.
nucleotide. A molecule composed of phosphate, 5-carbon sugar (ribose or deoxyribose), and a nitrogenous base.
nucleus. In a eukaryotic cell, the large organelle that contains the DNA; in an atom, the positively charged central portion that constitutes nearly all of the atomic mass and consists of protons and neutrons. *See* mass number.

ontogeny. The growth and development of an individual from fertilized egg to sexual maturity and senescence. *Compare* embryology, phylogeny.
organelle. A discrete structure within a cell; an organ of the cell.
organic chemistry. The study of the chemistry of materials derived from living matter, such as amber or petroleum. *Compare* biochemistry.
ornithology. The scientific study of birds.

pair bonding. A close and long-lasting relationship between a male and a female of a species based typically on food sharing and sex, usually for the cooperative rearing of young. *See* monogamy.
paleoanthropology. The scientific study of the fossil and cultural remains of ancient forms of hominid life. *Compare* anthropology, archaeology, primatology.
paleontology. The scientific study of past animal and plant life. *Compare* archaeology.
paradigm. A pattern, example, or model; in science, the prevailing view of a broad field of study.
parasitism. The living together of two unlike organisms such that one (the parasite) derives benefit and the other (the host) suffers illness or death.
parsimony. The principle of simplicity of assumptions (the simplest explanation is often found to be the best one).
photosynthesis. The chemical process whereby green plants make organic compounds from carbon dioxide and water in the presence of sunlight, chiefly in the leaves or needles. *See* chlorophyll, chloroplast.

phylogenetic lineage. *See* lineage.

phylogeny. The evolutionary history of a lineage. *Compare* ontogeny.

physiology. The scientific study of the functional and biochemical processes in organisms. *Compare* morphology.

polyandry. The practice or tendency of a female to mate with more than one male during one period of time. *See* polygamy.

polygamy. A life cycle that involves having more than one mate, whether serially or concurrently. *See* polyandry, polygyny.

polygyny. The practice or tendency of a male to mate with more than one female during one period of time. *See* polygamy.

polymorphism. Having several very different forms. *See* balanced polymorphism, dimorphism.

polypeptide. A molecule composed of three or more amino acids joined in a chain.

polyploidy. A condition in which an organism's haploid number of chromosomes is multiplied three, four, five, or more times. *Compare* allopolyploidy.

population. All of the individuals of a species or subspecies occupying a given area and sharing a common gene pool; a colony.

preadaptation. A preexisting condition that predisposes an organism for survival in an environment other than that in which it typically occurs.

prebiotic chemistry. Chemistry that existed before the evolution of life. *Compare* biochemistry.

prehistoric. Occurring in the time prior to the invention of writing, thus at any time prior to about 5,000 years ago. *Compare* extinction.

primate. Any animal of the order of mammals to which lemurs, monkeys, apes, and humans belong, characterized by large cerebral hemispheres, forward-directed eyes, opposable thumbs and/or big toes, and a lengthy developmental period.

primatology. The scientific study of primates, generally excluding recent man. *Compare* anthropology, archaeology, paleoanthropology.

primitive. Sharing the characteristics of the ancestral group. *Compare* derived.

prokaryote. A single-celled organism without a nuclear membrane, such as bacteria and cyanobacteria. *Compare* eukaryote.

protein. One of a group of large organic compounds composed of chains of amino acids.

protocell. A stage in the evolution of cells preceding the evolution of a coding mechanism such as DNA.

proton. A positively charged elementary particle present in the nucleus of every atom that is approximately equal in mass to the neutron. *See* mass number, nucleus, neutron.

punctuated equilibrium. A hypothesized mode of evolution in which species remain stable for long periods of time (equilibrium) and then change rapidly (punctuation). *Compare* gradualism, saltation.

radial symmetry. The arrangement of an animal's body parts around a central axis so that any plane cut at right angles through the axis results in halves that are mirror images of one another, as in jellyfish. *Compare* bilateral symmetry.

radiation. *See* adaptive radiation.

radioactive decay. The spontaneous transformation of an unstable atomic nucleus to a stable one.

radiometric dating. Any of a variety of techniques for determining the absolute age of a sediment or fossil based upon the fact that naturally occurring radioisotopes decay at a known rate. *See* absolute age.

red shift. Displacement of the lines in the spectra of galaxies to the longer wavelengths (red end of spectrum) as the galaxies recede from the observer.

regulator gene. A gene that controls the expression of the genes that control protein synthesis.

relative age. The geologic age of a fossil or rock defined relative to other fossils or rocks rather than in years. *Compare* absolute age.

reproductive isolation. The inability of two species, for any of a variety of reasons, to produce viable, fertile offspring. *See* behavioral, gametic, mechanical, seasonal isolation.

saltation. A mutation of great magnitude and rapid effect; a jump. *Compare* punctuated equilibrium.

scientific method. The commonsense but rigorous process by which one seeks to gain new knowledge or explain natural phenomena on the basis of observation, statement of a problem, formation of a hypothesis, experimentation or prediction to test the hypothesis, and, eventually, formation of a theory.

seasonal isolation. Reproductive isolation between two species owing to their coming into reproductive condition during different breeding seasons.

sediment. Material deposited by wind, water, or ice, eventually compressed into rock strata.

segment. *See* chromosome segment.

selection. *See* natural selection.

selective breeding. *See* artificial selection; *compare* natural selection.

senescence. The process of becoming old. *See* ontogeny.

serum. The liquid portion of the blood after clotting.

sex cell. The haploid product of meiosis; a sperm or egg: a gamete.

sexual dimorphism. Having marked anatomical differences between males and females of the same species, aside from the difference in sex organs. *Compare* polymorphism.

sexual reproduction. Reproduction whereby two gametes (haploid sex cells: one female, the other male) fuse to produce a zygote (diploid cell) that develops into a new organism. *Compare* asexual reproduction.

speciation. The evolutionary process yielding the formation of new species, usually requiring genetic variation and geographical isolation.

species (sing. and pl.). A group of interbreeding or potentially interbreeding organisms reproductively isolated from other such groups; the basic category of the classification of living or extinct organisms. *See* subspecies.

sperm. The mature, mobile, male gamete of sexually reproducing organisms. *Compare* egg.

spontaneous generation. The origin of life from non-living material via natural processes; abiogenesis; emergent evolution.

stabilizing selection. Selection for a well-adapted mean, thereby tending to eliminate the poorly adapted variants.

stasis. A period of evolutionary stability with little or no change. *Compare* equilibrium.

sterility. The inability of an individual organism to reproduce; failure to produce viable gametes.

stratigrapher. A geologist who studies rock strata.

stratigraphic sequence. The chronological succession of sedimentary rocks from older underneath to younger above.

stratigraphy. The study of geological layers formed by materials deposited by wind or water; the study of rock strata.

stratum, pl. **strata.** A layer of sedimentary rock.

subspecies. A geographical race; a localized population of a species that differs genetically, morphologically, and taxonomically from other populations of the species, but can interbreed with them if in contact.

superposition. An arrangement among geological strata such that the oldest stratum in a sequence is at the bottom and the youngest is at the top.

symbiosis. The living together of two unlike organisms in close association, with or without benefit or injury to either. *See* parasitism, mutualism, commensalism.

systematics. The study of the evolutionary relationships of organisms, practiced by systematists; often considered synonymous with taxonomy.

taxonomy. The classification of organisms into hierarchical groups—species into genera, genera into families, etc.—as practiced by taxonomists; often considered synonymous with systematics.

terrestrial. Living or moving about on land. *Compare* arboreal.

theory. A general principle describing processes or conditions in nature that is supported by a substantial body of evidence and has been repeatedly tested via the scientific method and found applicable in a wide variety of circumstances. *Compare* hypothesis, scientific method.

transitional fossil. A fossil that shares features with two different taxonomic groups, reflecting the probable evolution of one group from the other; a so-called missing link.

translocation. The exchange of segments between chromosomes.

unstable. The characteristic condition of radioactive substances that decompose into new forms spontaneously, predictably, and continuously; radioactive.

vestige. The degenerate remains, in a species, of an anatomical structure that was once more substantial, more functional, and of greater importance in an ancestral species.

vestigial. Degenerate or undeveloped; reduced in size and function by natural selection as no longer required, such as the hind-limb girdle of pythons and boa constrictors that shows that snakes evolved from lizards with legs. *Compare* derived, primitive.

virus. A noncellular infectious particle of nucleic acid, surrounded by a protein coat, that can reproduce only inside a host cell.

zygote. A fertilized egg.

Further Reading

Some books and articles well worth consultation are listed here in three groupings: the process of evolution, responses to creationism, and creationist publications. Many of these works are cited in the text.

The Process of Evolution

For the reader who wants more information, the following are a few of the most up-to-date scientific references on topics discussed in this book:

Asimov, I. 1989. *Beginnings: The Story of Origins—of Mankind, Life, the Earth, the Universe*. Berkeley Books, New York. 289 pp. (An entertaining, easy-to-read paperback on origins.)

Attenborough, D. 1979. *Life on Earth*. Little, Brown and Co., Boston. 319 pp. (Well-illustrated general natural history that parallels the excellent PBS television series of the same name.)

Ayala, F. J. 1982. *Population and Evolutionary Genetics: A Primer*. Benjamin/Cummings Publ. Co., Menlo Park, Cal. 268 pp. (Deals with the evolutionary process at the genetic level.)

Berra, T. M. 1980. Charles Darwin: What else did he write? *Amer. Biol. Teacher* 42(8): 489–92. (List of books by and about Darwin.)

Caccone, A., and J. R. Powell. 1989. DNA divergence among hominoids. *Evolution* 43: 925–42.

Campbell, B. G. 1985. *Humankind Emerging*, 4th. ed. Little, Brown and Co., Boston. 528 pp. (Comprehensive college-level paleoanthropology text.)

Ciochon, R. L., and R. S. Corruccini (eds.). 1983. *New Interpretations of Ape and Human Ancestry*. Plenum Press, New York. 888 pp. (Excellent

anthology of highly technical arguments by leading anthropologists.)
Colbert, E. H. 1980. *Evolution of the Vertebrates*, 3rd ed. Wiley, New York. 510 pp. (Standard reference on the history of backboned animals through time, very well illustrated.)

Dalrymple, G. B. *The Age of the Earth*. Stanford University Press, Stanford, Calif. About 550 pp. In press. (A definitive history, clearly written, of what people and science have thought and learned about the age of the Earth; an abridged edition will also appear.)

Dawkins, R. 1976. *The Selfish Gene*. Oxford Univ. Press, New York. 224 pp. (Explains how behaviors such as reciprocal altruism may have evolved.)

Dawkins, R. 1986. *The Blind Watchmaker*. W. W. Norton & Co., New York and London. 332 pp. (Very thoughtful book on the cumulative effects of natural selection that make a designer unnecessary.)

Day, M. H. 1986. *Guide to Fossil Man*, 4th ed. Univ. Chicago Press, Chicago. 43 pp. (The standard reference on hominid remains.)

Delson, E. 1987. Evolution and palaeobiology of robust *Australopithecus*. *Nature* 327: 654–55.

Dobzhansky, T., F. J. Ayala, G. L. Stebbins, and J. W. Valentine. 1977. *Evolution*. Freeman, San Francisco. 572 pp. (A very good university-level evolution text and reference.)

Dodson, E. O., and P. Dodson, 1985. *Evolution: Process and Product*, 3rd ed. Wadsworth Publ. Co., Belmont, Cal. 596 pp. (A current university-level evolution text and reference.)

Dott, R. H., Jr., and R. L. Batten. 1981. *Evolution of the Earth*, 3rd. ed. McGraw-Hill, New York. 573 pp. (University-level textbook in historical geology.)

Edey, M. A., and D. C. Johanson. 1989. *Blueprints: Solving the Mystery of Evolution*. Little, Brown, Boston. 417 pp. (An absorbing history of the evolutionary idea.)

Eldredge, N., and S. J. Gould. 1972. Punctuated equilibria: An alternative to phyletic gradualism. Pp. 82–115 *in* T. J. M. Schopf (ed.), *Models in Paleobiology*. Freeman, Cooper, and Co., San Francisco. (The title tells the story.)

Eldredge, N., and S. M. Stanley. 1984. *Living Fossils*. Springer Verlag, New York. 291 pp. (Discusses 34 examples of arrested evolution.)

Eldredge, N., and I. Tattersall. 1982. *The Myths of Human Evolution*. Columbia Univ. Press, New York. 197 pp. (Human evolution from a punctuated-equilibrium point of view; attacks gradualism.)

Fabian, A. C. (ed.). 1988. *Origins*. Cambridge University Press, Cambridge, England. 168 pp. (Collection of technical articles by experts on origins, from the Big Bang to humans and our behavior, society, and language.)

Feduccia, A. 1980. *The Age of Birds*. Harvard Univ. Press, Cambridge, Mass. 196 pp. (Account of avian evolution, with excellent illustrations.)

Fleagle, J. G. 1988. *Primate Adaptation and Evolution*. Academic Press, San Diego. 486 pp. (The most recent textbook on primate evolution.)

Fox, S. W., and K. Dose. 1977. *Molecular Evolution and the Origin of Life*, rev. ed. Dekker, New York. (Highly technical; gives recipe for the synthesis of protocells. For an easy-to-read summary, see S. Fox, 1981. From inanimate matter to living systems. *American Biol. Teacher* 43(3): 127–40.)

Futuyma, D. J. 1986. *Evolutionary Biology*, 2nd ed. Sinauer Associates, Inc., Sunderland, Mass. 600 pp. (A current university-level evolution text and reference.)

Gould, S. J. 1977. *Ever Since Darwin*. W. W. Norton Co., New York. 285 pp.

Gould, S. J. 1980. *The Panda's Thumb*. W. W. Norton Co., New York. 243 pp.

Gould, S. J. 1983. *Hen's Teeth and Horse's Toes*. W. W. Norton Co., New York. 413 pp.

Gould, S. J. 1985. *The Flamingo's Smile*. W. W. Norton Co., New York. 476 pp. (The four books by Gould are collections of delightful, thought-provoking essays on various evolutionary topics.)

Grant, P. R. 1986. *Ecology and Evolution of Darwin's Finches*. Princeton University Press, Princeton, N.J. 458 pp. (The most recent study and review of these fascinating birds.)

Gribbin, J., and J. Cherfas. 1982. *The Monkey Puzzle*. McGraw-Hill, New York. 280 pp. (An attempt to explain molecular anthropology to the general reader.)

Hamblin, W. K. 1982. *The Earth's Dynamic Systems*, 3rd. ed. Burgess, Minneapolis. 529 pp. (A recent university-level text in physical geology.)

Harland, W. B., A. V. Cox, P. G. Llewellyn, C. A. G. Pickton, A. G. Smith, and R. Walters. 1982. *A Geologic Time Scale*. Cambridge Univ. Press, Cambridge, England. 131 pp. (The latest revision of geologic time; highly technical.)

Harrison, E. R. 1981. *Cosmology*. Cambridge Univ. Press, Cambridge, England. 430 pp. (A technical, but readable, introduction to the science of the universe.)

Hickman, C. P., Jr., L. S. Roberts, and F. M. Hickman. 1988. *Integrated Principles of Zoology*, 8th ed. Mosby, St. Louis. 939 pp. (One of the best university-level zoology texts and reference books.)

Johanson, D., and M. Edey. 1981. *Lucy: The Beginnings of Humankind*. Simon & Schuster, New York. 409 pp. (Highly readable account of the behind-the-scenes operation of paleoanthropologists and information on human origins.)

Johanson, D. C., et al. 1987. New partial skeleton of *Homo habilis* from Olduvai Gorge, Tanzania. *Nature* 327: 205–9. (Also see pp. 187–88.)

Kimbel, W. H., T. D. White, and D. C. Johanson. 1988. Implications of KNM-WT 17000 for the evolution of "robust" *Australopithecus*. Pp. 259–68 *in* F. E. Grine (ed.), *Evolutionary History of "Robust" Australopithecines*. Aldine de Gruyter, New York.

Leakey, R. E. 1981. *The Making of Mankind*. S. P. Dutton, New York. 256 pp. (Beautifully illustrated account of human evolution by an author who ought to know.)

Lewin, R. 1982. *Thread of Life*. Smithsonian Books, Washington, D.C. 256 pp. (Fine coffee-table book for the layman.)

Lewin, R. 1984. *Human Evolution*. W. H. Freeman, New York. 104 pp. (An excellent, well-illustrated, easily read introduction.)

Lewin, R. 1987. *Bones of Contention*. Simon & Schuster, New York. 348 pp. (A fascinating inside look at quarrels among paleoanthropologists.)

Lewin, R. 1988. *In the Age of Mankind*. Smithsonian Books, Washington, D.C. 255 pp. (Superbly illustrated and highly readable guide to the latest information on human evolution.)

Lovejoy, C. O. 1981. The origin of man. *Science* 211: 341–50. (Readable, thought-provoking essay on the origin of bipedality. For opposing views, see letters to editor in *Science* 217: 295–306.)

Mayr, E. 1982. *The Growth of Biological Thought*. Belknap Press, Cambridge, Mass. 974 pp. (A magnum opus on diversity, evolution, and inheritance by a leading figure in evolutionary biology.)

Miller, J., and B. van Loon. 1982. *Darwin for Beginners*. Pantheon Books, New York. 176 pp. (Clever, cartoon-illustrated, but scientifically correct explanation of Darwinism, the sort of book students could give to their parents to help them understand evolution.)

Otte, D., and J. A. Endler (eds.). 1989. *Speciation and Its Consequences*. Sinauer Associates, Sunderland, Mass. 679 pp. (A compendium of technical articles on species formation.)

Oxnard, C. E. 1987. *Fossils, Teeth and Sex*. Univ. Washington Press, Seattle and London. 281 pp. (Recent study of the implications of sexual dimorphism.)

Pilbeam, D. 1984. The descent of hominoids and hominids. *Scientific American* 250 (3): 84–96. (Well-illustrated summary as of 1983.)

Raup, D. M., and S. M. Stanley. 1978. *Principles of Paleontology*, 2nd ed. Freeman, San Francisco. 481 pp. (Modern university-level text.)

Reader, J. 1986. *The Rise of Life: The First 3.5 Billion Years*. Collins, London.

192 pp. (The history of life as a detective story, illustrated with the author's beautiful photographs of historical specimens, manuscripts, and artifacts.)

Romer, A. S., and T. S. Parsons. 1977. *The Vertebrate Body*. Saunders, Philadelphia. 624 pp. (Well-illustrated classic of comparative anatomy.)

Ruse, M. 1986. *Taking Darwin Seriously*. Basil Blackwell, Oxford and New York. 303 pp. (Explains how evolutionary processes can account for the origin of morality.)

Scientific American. 1978. *Evolution*. Sept.: 239(3). (Nine articles by leading experts on various aspects of evolution.)

Shackley, M. 1980. *Neanderthal Man*. Archon Books, Hamden, Conn. 149 pp. (Restores the dignity of Neanderthals, once lost to bad movies.)

Shipman, P. 1986. Baffling limb on the family tree. *Discover*. Sept.: 87–93. (Also see Johanson's comments in Nov., pp. 116–17.)

Sibley, C. G., and J. E. Ahlquist, 1987. DNA hybridization evidence of hominoid phylogeny: Results from an expanded data set. *Journal of Molecular Evolution* 26: 99–121. (See comments on this technique in *The Scientist*, June 13, 1988; and in *Science* 241: 1756–59, 1988; 242: 651–52, 1624.)

Simons, E. 1989. Human origins. *Science* 245: 1343–50. (A review of the most recent developments in paleoanthropology.)

Smith, J. M. (ed.). 1982. *Evolution Now: A Century after Darwin*. W. H. Freeman & Co., San Francisco. 239 pp. (Twenty-one recent controversial topics by well-known authorities, with explanatory comments by J. M. Smith.)

Stanley, S. M. 1981. *The New Evolutionary Timetable*. Basic Books, New York. 222 pp. (Popular account of punctuated equilibrium.)

Stebbins, G. L. 1982. *Darwin to DNA, Molecules to Humanity*. W. H. Freeman, San Francisco. 491 pp. (Review by a leading geneticist.)

Stringer, C. 1988. The dates of Eden. *Nature* 331: 565. (Comments on relationship of Neanderthals and modern humans in light of Valladas et al., 1988.)

Susman, R. L. 1988. Hand of *Paranthropus robustus* from Member 1, Swartkrans: Fossil evidence for tool behavior. *Science* 240: 781–84. (See also R. Lewin's commentary on pp. 724–25 of the same issue.)

Tattersall, I., and E. Delson. 1984. *Ancestors: Four Million Years of Humanity*. American Museum of Natural History, New York. 12 pp. (Guidebook to a wonderful exhibition at the American Museum, 13 April to 9 September, 1984.)

Valladas, H., J. L. Reyss, J. L. Joron, G. Valladas, O. Bar-Yosef, and B. Vandermeersch. 1988. Thermoluminescence dating of Mousterian

Proto-Cro-Magnon remains from Israel and the origin of modern man. *Nature* 331: 614–16.

Walker, A., R. E. Leakey, J. M. Harris, and F. H. Brown. 1986. 2.5 Myr *Australopithecus boisei* from west of Lake Turkana, Kenya. *Nature* 322: 517–22. (Also see pp. 496–97.)

Weaver, K. F., D. L. Brill, and J. H. Matternes. 1985. The search for our ancestors. *National Geographic* 168(5): 560–623, November. (Excellent presentation of human evolution, with holographic image of Taung child on cover.)

Wilson, A. C. 1985. The molecular basis of evolution. *Scientific American* 253(4): 164–73.

Wilson, E. O. 1975. *Sociobiology*. Harvard Univ. Press, Cambridge, Mass. 697 pp. (This landmark work extends evolution into the field of animal behavior.)

Zihlman, A. L. 1982. *The Human Evolution Coloring Book*. Barnes and Noble Books, New York. 124 pp. (Well-illustrated beginner's guide to a variety of topics.)

Responses to Creationism

Creationists have succeeded in provoking scientists and other thinkers into explaining the dangers of creationism and why it is unscientific. The following recent books address those issues. The titles are self-explanatory and most are available in paperback.

American Society of Zoologists. 1984. Science as a way of knowing. *American Zoologist* 24(2): 421–534.

Beck, S. D. 1982. Natural science and creation theology. *Bioscience* 32: 738–42.

Charig, A. J., F. Greenaway, A. C. Milner, C. A. Walkey, and P. J. Whybrow. 1986. *Archaeopteryx* is not a forgery. *Science* 232: 622–26.

Creation/Evolution. (This timely periodical answers the arguments raised by creationists. It is published quarterly and is available from P.O. Box 5, Amherst Branch, Buffalo, N.Y. 14226. The volumes published so far are extremely useful to teachers.)

Creation/Evolution Newsletter. (This bimonthly focuses on current events, debates, and discussion. It is dedicated to promoting and defending the integrity of science education and is a nice adjunct to *Creation/Evolution*. The *Creation/Evolution Newsletter*, now known as *NCSE Reports*, is available from Box 9477, Berkeley, Cal. 94709.)

Eldredge, N. 1982. *Monkey Business: A Scientist Looks at Creationism*. Washington Square Press, New York. 157 pp.

Eve, R. R., and F. B. Harrold. 1986. Creationism, cult archaeology and other pseudoscientific beliefs: A study of college students. *Youth and Society* 17(4).

Fuerst, P. A. 1984. University student understanding of evolutionary biology's place in the creation/evolution controversy. *Ohio Journal of Science* 84(5): 218–28. (A shameful revelation.)

Futuyma, D. J. 1983. *Science on Trial: The Case for Evolution*. Pantheon Books, New York. 251 pp. (The best of this genre.)

Godfrey, L. R. (ed.). 1983. *Scientists Confront Creationism*. W. W. Norton & Co., New York. 324 pp.

Harrold, F. B., and R. A. Eve. 1986. Noah's Ark and ancient astronauts: Pseudoscientific beliefs about the past among a sample of college students. *Skeptical Inquirer* 11: 61–75.

Harrold, F. B., and R. A. Eve (eds.). 1987. *Cult Archaeology and Creationism*. Univ. Iowa Press, Iowa City. 175 pp. (More depressing surveys documenting the decline of science education.)

Journal of Geological Education. 1982. Vol. 30(1). (This issue is devoted to answering creationists' arguments about the age of the Earth. Vol. 31, pp. 72–134, also refutes creationist claims. See later issues as well.)

Kitcher, P. 1982. *Abusing Science: The Case against Creationism*. MIT Press, Cambridge, Mass. 213 pp.

LaFollette, M. C. (ed.). 1983. *Creationism Science and the Law: The Arkansas Case*. MIT Press, Cambridge, Mass. 236 pp.

Lazar, E. 1987. *Creation/Evolution Bibliography/Directory*, 3rd. ed. Published by author, 1324 G St. #4, Sacramento, Cal. 95814-1543. (Useful looseleaf compilation of source material such as dissertations, newspaper articles, legal opinions, debates, etc.)

Mather, K. F. 1982. The Scopes trial and its aftermath. *Journal of the Tennessee Academy of Science*. 57(1): 1–9.

McCollister, B., ed. 1989. *Voices for Evolution*. National Center for Science Education, Berkeley, Calif. 141 pp. (This compendium lists the statements of 86 scientific, religious, and educational organizations in support of evolution and in opposition to creationism.)

Montagu, A. (ed.). 1984. *Science and Creationism*. Oxford Univ. Press, New York. 320 pp.

NCSE Reports. (See *Creation /Evolution*, above, the former name of these reports.)

Newell, N. D. 1982. *Creation and Evolution: Myth or Reality?* Columbia Univ. Press, New York. 203 pp.

Numbers, R. L. 1982. Creationism in twentieth-century America. *Science* 218: 538–44.

Overton, W. R. 1982. Creationism in schools: The decision in McLean versus the Arkansas Board of Education. *Science* 215: 934–43. (This complete text of Judge Overton's judgment, injunction, and opinion reestablishes one's confidence in the American judicial system.)

Ruse, M. 1982. *Darwinism Defended*. Addison-Wesley, Reading, Mass. 356 pp. (A guide to evolution controversies that persuasively shows why Darwinism is such a powerful explanatory tool for biologists.)

Scott, E. C., and H. P. Cole, 1985. The elusive scientific basis of creation "science." *Quarterly Review of Biology* 60(1): 21–30. (Documents the lack of scientific articles by creationists.)

Strahler, A. N. 1987. *Science and Earth History: The Evolution/Creation Controversy*. Prometheus Books, Buffalo, N.Y. 552 pp. (A comprehensive critique of creationism by a highly qualified geologist.)

Weinberg, S. (ed.). 1984. *Reviews of Thirty-One Creationist Books*. National Center for Science Education, Inc., Berkeley, Calif.

Wilson, D. B. (ed.). 1983. *Did the Devil Make Darwin Do It?* Iowa State Univ. Press, Ames. 242 pp.

Zimmerman, M. 1986. The evolution-creation controversy: Opinions from students at a "liberal" liberal arts college. *Ohio Journal of Science* 86(1): 134–39.

Zimmerman, M. 1987. The evolution-creation controversy: Opinions of Ohio high school teachers. *Ohio Journal of Science* 87(4): 115–24.

Zimmerman, M. 1988. Ohio school board presidents' views on the evolution-creation controversy: Part 2. *Newsletter of the Ohio Center for Science Education*, January 1988.

Creationist Publications

The following is a list of the major publications in which creationists speak for themselves. Reviews of creationist publications can be found in S. Weinberg (ed.). 1984. *Reviews of Thirty-one Creationist Books*. National Center for Science Education, Inc., Syosset, N.Y.

Gish, D. T. 1979. *Evolution? The Fossils Say No!* Creation-Life Publishers, San Diego. 189 pp.

Moore, J. N., and H. S. Schultz (eds.). 1970. *Biology: A Search for Order in Complexity*. Zondervan Publishing House, Grand Rapids, Mich.

Morris, H. M. 1974. *Scientific Creationism*. Creation-Life Publishers, San Diego. 277 pp.

Morris, H. M. 1974. *The Troubled Water of Evolution*. Creation-Life Publishers, San Diego.

Morris, H. M. 1978. *The Remarkable Birth of Planet Earth*. Creation-Life Publishers, San Diego.

Morris, H. M. 1984. *A History of Modern Creationism*. Master Book Publishers, San Diego.
Slusher, H. S., and T. P. Gramwell. 1978. *The Age of the Earth*. Creation-Life Publishers, San Diego.
Whitcomb, J. C., Jr., and H. M. Morris. 1961. *Genesis Flood*. Presbyterian and Reformed Publishing Co., Nutley, N.J.

Index

About the Author

Dr. Tim M. Berra is Professor of Zoology at The Ohio State University at Mansfield and a two-time Fulbright fellow to Australia. He is the former Editor of the *Ohio Journal of Science*, and is the author of two other books and over forty scientific papers. Dr. Berra received his Ph.D. in biology from Tulane University in 1969. He is a broadly based vertebrate zoologist with special interests in fishes, zoogeography, ecology, Australian natural history, and human evolution.

Library of Congress Cataloging-in-Publication Data

Berra, Tim M., 1943–
Evolution and the myth of creationism : a basic guide to the facts in the evolution debate / Tim M. Berra.
p. cm.
Includes bibliographical references.
ISBN 0-8047-1548-3. — ISBN 0-8047-1770-2 (pbk.)
1. Evolution. 2. Creationism—Controversial literature.
3. Creationism—Study and teaching—United States. 4. Creationism—Study and teaching—Law and legislation—United States. I. Title.
QH371.B47 1990 89-51484
213—dc20 CIP

This book is printed on acid-free paper.